THE BOOK OF PEARS AND PLUMS

LECTOR HOUSE PUBLIC DOMAIN WORKS

This book is a result of an effort made by Lector House towards making a contribution to the preservation and repair of original classic literature. The original text is in the public domain in the United States of America, and possibly other countries depending upon their specific copyright laws.

In an attempt to preserve, improve and recreate the original content, certain conventional norms with regard to typographical mistakes, hyphenations, punctuations and/or other related subject matters, have been corrected upon our consideration. However, few such imperfections might not have been rectified as they were inherited and preserved from the original content to maintain the authenticity and construct, relevant to the work. The work might contain a few prior copyright references as well as page references which have been retained, wherever considered relevant to the part of the construct. We believe that this work holds historical, cultural and/or intellectual importance in the literary works community, therefore despite the oddities, we accounted the work for print as a part of our continuing effort towards preservation of literary work and our contribution towards the development of the society as a whole, driven by our beliefs.

We are grateful to our readers for putting their faith in us and accepting our imperfections with regard to preservation of the historical content. We shall strive hard to meet up to the expectations to improve further to provide an enriching reading experience.

Though, we conduct extensive research in ascertaining the status of copyright before redeveloping a version of the content, in rare cases, a classic work might be incorrectly marked as not-in-copyright. In such cases, if you are the copyright holder, then kindly contact us or write to us, and we shall get back to you with an immediate course of action.

HAPPY READING!

THE BOOK OF PEARS AND PLUMS

EDWARD BARTRUM,
HARRY ROBERTS

ISBN: 978-93-93693-79-2

Published: -

LECTOR HOUSE

LECTOR HOUSE LLP
E-MAIL: lectorpublishing@gmail.com

Pear Blossom

(From a drawing by Ethel Roskruge)

HANDBOOKS OF PRACTICAL GARDENING—XI

THE BOOK OF PEARS AND PLUMS

BY

THE REV. E. BARTRUM, D.D., F.R.H.S.

RECTOR OF WAKES COLNE, ESSEX
EDITOR OF "HELPFUL HINTS FOR HARD TIMES," ETC.

WITH CHAPTERS ON CHERRIES AND MULBERRIES

EDITED BY
HARRY ROBERTS

CONTENTS

PLUMS AND DAMSONS

CHERRIES

THE MULBERRY

APPENDIX

LIST OF ILLUSTRATIONS

Page

INTRODUCTION

I have grown pears, plums, cherries and mulberries for many years, and have written many articles about the first two fruits; yet, in preparing this work, I found that I had still much to learn, and I wish particularly to express my obligations to the new edition of Thompson's *Gardener's Assistant,* edited in six volumes by Mr Watson, Assistant Curator of the Royal Gardens, Kew, and brought out by the Gresham Publishing Company. I have also derived valuable aid from the volumes of the Royal Horticultural Society. The chapter on "cherries" is based chiefly on the booklet contributed by Mr G. Bunyard to my *Helpful Hints for Hard Times* published by the S.P.C.K.

E. B.

WAKES COLNE RECTORY, ESSEX,
July 1902.

THE BOOK OF PEARS AND PLUMS

PEARS

HISTORY OF THE PEAR

The Pear is my theme, and a pleasant one it is. Only those who have planted trees, pruned them, watched their growth, plucked the fruits, enjoyed them at almost all hours, seen them on the table month after month as an appetising dish, can fully realise the value of the Pear. A good Pear-tree is like a faithful friend—treat him properly and he will not fail you. Circumstances, as for instance, a late frost, may render him incapable of helping you; he may have nothing to offer you; no doubt he is sorry, but with patience he will do you a good turn.

Pyrus (or *pirus*), the Latin name for Pear-tree, is the name of a genus of plants belonging to the natural order Rosaceæ. *Pyrus communis*, the wild pear, from which the numerous cultivated varieties have sprung, is found over a great part of Europe and Asia, within the limits of the temperate regions. Its origin is lost in obscurity. The lake-dwellers in Switzerland are said to have stored the fruits for winter use. It was probably brought by the Greeks, possibly by birds, from Asia, and after a time became a favourite with the Romans as well as the Greeks. It is mentioned by Horace, Vergil, Juvenal, and others. Pliny refers to numerous varieties, describing those with special flavours. He tells us that many of the sorts were called after the countries from which they came, such as the Syrian, the Alexandrian, the Numidian, and the Grecian. Thus he mentions *pira nardina*, a pear with the scent of nard; *pira onynchina*, a pear of the colour of the fingernail, and others. These last are evidently Greek. Forty or fifty sorts are named in Roman writers, and the Pear was appropriately dedicated to Minerva, the goddess of wisdom.

The Romans no doubt took their pear-trees northwards into Gaul and Britain. The climate of France is so well adapted to the growth of pears, that at one time it was thought all good pears must come from France. I well remember many years ago seeing a garden in this country full of pear-trees, every one of which had come from France. Happily there is no need now to go out of England for the very best varieties. A list published in 1628 by a fruit-grower of Orleans named Le Lectier (there is a new variety called by his name, and probably after him) enumerates 260 varieties. The well known Jargonelle is mentioned in that list. Our Parkinson in 1629 refers to 64 varieties only. Seventy years later we read of 138, and in 1829 of 630 varieties. John Scott, rather famous as a fruit-grower forty years ago, says in his "Orchardist" that he has above 1000 sorts worked upon the Quince Stock.

He had studied pomology at the "*Jardin Fruitier*," the fruit garden attached to the *Jardin des Plantes* at Paris, and, using his opportunities, learnt all the secrets of Pear culture, and brought them from France to Merriott, near Crewkerne, in Somerset. The last edition of Dr Hogg's "Fruit Manual" (invaluable to the Pomologist), published in 1884, contains the names of 647 varieties. Not a few of these were marked as worthless by the Committee of the National Pear Conference, held at Chiswick in October 1885. The Royal Horticultural Society in their "Fruits for Cottagers and Small Farmers" (1892), selected eight varieties only for eating, and two more for late keepers; four were recommended for cooking or stewing. Fresh sorts are constantly being brought into notice, the result of cross-fertilisation, and we may, I think, congratulate ourselves that British pears in a favourable season are as good as those produced in any part of the world. Let any one who doubts this statement attend a Crystal Palace or any other first-rate Fruit Show; his doubts will soon be dispelled.

SITUATION AND SOIL

These two points are of the greatest importance in successful cultivation. No amount of skill will enable even a clever gardener to grow good fruit in a bad site. Where the land is low and swampy, exposed therefore to frosts more than ground at a higher altitude, the effort would be useless. Stagnant water moreover produces canker, and soon ruins trees. Pears love a deep moist soil, but not water that lies for any length of time about the roots. On a hillside, where the slope is more than gradual, so that in a dry season the upper part suffers from drought, they would be a failure. Trees planted near the bottom and properly protected from winds might succeed, yet they would probably suffer from frost. The slope should not be more than two to three feet in a hundred.

The aspect should be south, south-east or south-west. The Pear is of Eastern origin, and probably retains its Eastern habit in blooming early some time before the apple. It needs more warmth, and more protection.

To plant pears in a north aspect even on a wall is a mistake. Morello cherries are a sure crop, pears a very doubtful one. The wood is not well ripened, and bloom-buds are not often formed. The amount of rainfall is also a matter for consideration. If the soil is light, more moisture will be needed than in heavier land. Heavy clays are not good for pears, yet much may be done to improve such soils, and some outlay may be desirable in gardens and small plantations. Good drainage will be necessary. The ground before planting must be well lifted and exposed to the air; some portions should be burnt and mixed with the rest; decayed vegetable matter should be added in abundance. After planting, when the trees are rooted and growing, the soil should be often lifted with a light fork, or hoed, and the air admitted to the roots. A clayey loam is the best of all soils for the Pear, yet even that may be much improved by exposure before planting, and the use of the fork or hoe afterwards. In sandy or chalky soils, pears will have a poor chance even on the free (or pear) stock, unless the ground has been previously prepared by trenching, and then digging in a good quantity of decayed stable or farmyard manure. Marl or clay from other parts, or turf (chopped up) from a field, may be added with advantage. Generous treatment subsequently in the way of liquid manure will alone make trees in such ground a success. Should, however, the soil be shallow and the subsoil gravel or chalk, trees must be lifted every few years, and the expense in a large garden might be considerable.

The monks in olden days were wont to put slates or large stones below their trees before planting, to prevent the tap-root running into bad soil. In modern gardens a concrete bottom two or three inches thick, sloping towards a drain in front, is sometimes made. Methods must depend on soil and means. A concrete bottom is better than a stratum of stones or brick rubbish. Persons content with a few small trees may lift them frequently or root-prune annually, in which case no special precautions are required.

PROTECTION

As the Pear needs sunshine and warmth as well as moisture, it must have protection from cold winds. Walls and buildings are not always to be had. Black Italian or Canadian poplars well planted and rather close together soon form a good shelter; limes (invaluable for bees) quickly make a good fence if encouraged to throw shoots from the lower part of the tree and closely cut in. Hedges of damsons or the myrobalan (the cherry plum) serve as shelters from the wind and grow rapidly. This cherry plum blooms early, and its flower is often cut off; otherwise its fruit (ripe in August) is useful for tarts. Protection is needed on the south-west against the winds as well as on the north-east. The larger trees should be placed at some distance that their roots may not absorb the nourishment needed in the fruit garden.

PLANTING

seems a simple subject, yet the difference between good and bad work may make the difference between success and failure. Proper planting is of vital importance. The ground should be prepared beforehand. If it is wet, and the water does not readily pass off, drainage is essential. The depth of the drains must depend on the outfall. If they can be sunk three or even four feet below the surface, they are less exposed to danger from deep trenching or the roots above them. The drains should be about five yards apart. The soil should then be well trenched and exposed thoroughly to the action of the atmosphere. But beware of opening holes some time beforehand. Should rain come, the holes will be filled, and if the soil is heavy, may remain there for some time. Abstain, too, from planting in wet weather. If the ground is sticky, the roots will not have free play. Should the soil be light, well-decayed manure may be dug in, especially if it has been well mixed some time beforehand with turfy or good loam. In strong soil, no manure is needed. When the trees arrive, do not unpack them until you are ready to plant. Exposure of the roots to the air should be avoided as much as possible. If delay occurs from rain, frost, or any other cause, put the roots in the ground, laying the trees in a slanting position in a trench, and covering the roots thoroughly with soil. Choose, too, a sheltered position in the garden for the trench. Should the ground be hard from frost, do not unpack the trees; keep them under cover, and protect them as far as possible from cold and frost. When the ground is fit and the weather favourable, open the earth 2 to 3 feet across at a depth of 12 to 18 inches according to the class and size of the tree and roots. Carefully examine the roots. Cut off the points of any jagged or torn roots cleanly with a sharp knife, and shorten all downward and coarse roots. Cut on the under side, and towards the outside, so that the tree may lie flat. Avoid any injury to the rootlets. The aid of a lad will be useful to hold the tree in its place while the gardener is planting. Spread the roots and rootlets carefully out with an upward rather than a downward tendency. Then scatter fine soil amid them, shaking the trees occasionally, adding more soil until it stands erect. Now tread in the soil firmly, and fill up the hole with fresh soil, raising the earth several inches above the ordinary level. The soil will sink after a time, and occasionally more soil may be added subsequently. But deep planting should always be avoided.

With pears on the Quince, it is important that all the quince stock should be covered by the soil, as it suffers in dry weather if exposed, and the fruit would therefore be affected. All buds on this stock should on this account be inserted as near the ground as possible. Should the soil be very heavy, yet pears *must* be planted, place the roots almost on the surface, and throw the lightest earth obtainable round the stem. If such ground is trodden down hard, and rain should soon follow, the ground would probably become like a brick, and the roots, kept in check, would suffer seriously.

The best time for planting is towards the close of October and in November. Select your trees yourself, and go only to first rate nurserymen for pears if you want varieties on the Quince stock. Each nursery has its specialty. Budding, grafting and double-grafting on special stocks do not always have the attention and skill required. If you cannot go, send your orders early, so as to secure an early choice and good trees. Planting may continue to the end of February, but you must not expect good trees for late orders. The roots, too, make some progress even in winter, so that early planting is preferable in every way.

STAKING AND WIRING

Standards should be attached to a round, strong, stout stake 2½ inches thick or more, as soon as planted. The best plan is to have the stake in position ready for the tree. For full sized standards, the stakes should be 7 to 8 feet long, and driven 18 inches or more into the ground; they should be in the centre of each hole. Choose durable wood, as far as possible. A straw or hay band, or a piece of bagging, should now be run round the stem, and the stake attached to it by thick string or cord well tarred. The twigs of the willow (soft and strong, especially the golden willow) may also be used. Protection against rabbits must be provided at once. A wire fence round the orchard or garden is best; where there is no fence, put a yard of wire netting (1¼ mesh) round each tree. This will last for years. The wire should be 3 feet high at the least. Examine your fence every year in September and repair. You cannot be too particular. Serious damage may be done in a night.

STOCKS FOR PEARS

The discovery of the Quince Stock, as adapted to the Pear for budding or grafting upon, has added immensely of late years to the popularity of this valuable fruit. The discovery, it is true, is not a new one. Merlet, writing in 1667 (says Mr Scott), recommends the Portugal Quince as stronger and more favourable for working pears upon than any other variety: "It swells equally fast with the graft, which none of the other sorts do." Le Gendre, an author of about the same date, in *Le Manière de cultiver les arbres Frutiers*, says: "I have been much aided by the invention of grafting the Pear upon the Quince," and adds that he was one of the first who helped to introduce this method. By this discovery the well-known saying: "Plant pears for your heirs," must give way to another:—

"That those who plant pears
Grow fruit for their heirs
Is a maxim our grandfathers knew;
But folks have learnt since,
If you graft on the quince
The fruit will develop for you."[1]

This stock checks excessive growth, and brings the tree into early bearing. It is not adapted for large standards nor for light soil; in good pear ground it is simply invaluable. Sometimes poor results occur, but the failure is usually caused by the want of proper care, either at the nursery or in the garden. Young trees are often overworked. Some varieties will not thrive on the quince stock, so that double-grafting has been introduced. Thus the strong-growing Beurré d'Amanlis is grafted on the quince, then two years after some other sort is grafted on it. It is said that in this way Gansel's Bergamot is made "a marvel of fertility,"[2] but this is not my experience! The disappointing pear Marie Louise is usually double-grafted, so is that excellent late pear Josephine de Malines for cordons, bushes, or pyramids, and so are many others. Strong-growing varieties like Vicar of Winkfield, Beurré Hardy, Beurré Clairgeau, Marie Louise d'Uccle, and others, are used as intermediate stocks. To check the vigorous Pitmaston Duchess, the weakly Winter Nelis is employed as an intermediary. Our chief nurserymen are studying the habits of each pear which needs double grafting, and failure is rare on their part. Fruits grown on the Quince Stock are often more highly coloured, and not so coarse as such as are on the Pear Stock. Those who have a good pear soil then should plant no tree on the Pear Stock, except in an orchard.

The varieties usually employed are the Portugal, the Angers, and the common Quince. The Angers being compact, prolific, and easily increased, is said to be the favourite.[3]

In some soils Pear Stocks must be used. The Quince would not thrive; it is not strong enough. The latter is surface rooting, it emits more fibres, and does not rejoice in the tap-root of the Pear Stock. But for light and unfavourable soils, and

[1] *See* Cheal, "Fruit Culture," p. 8.

[2] Rivers.

[3] See an excellent article on Pears in new edition of Thompson's "Gardeners' Assistant," by R. L. C.

also for large standards, the Pear Stock alone will suffice. This is often called the Free Stock, as compared with the dwarfing Quince. In former years the seeds of the wild pear were used to raise new stocks, but at the present time pear seedlings are sent from France to England and the United States in large quantities. Our cousins, however, are exerting themselves earnestly to improve the pear, and with their energy and variety of climate, will not long be dependent upon France.

ORCHARD TREES

In good soil and a favourable, well-sheltered aspect, standard trees on the pear stock may be a success if planters and owners can wear the cap of patience for eight to ten years. Should it be probable that cattle will use the ground, a strong and lasting fence must be put round each tree, as thorns encircling them will not suffice. Iron fences made for the purpose, with wire netting added at the top, may be the cheapest in the end. Otherwise, put three posts (larch or oak) to form a triangle round the tree. These should be well charred or tarred at some distance from the lower end before being firmly driven in. The tops should slant outwards. Then nail cross-pieces to the posts; old railway-sleepers are sometimes cheap and useful. The standards in good soil should be thirty feet apart or more. It is a mistake to allow the grass at any time to grow under the trees. Moisture which pears require is absorbed, and the air is kept from the roots. Reduce the branches after planting (in October or November) to five or six at the most; cut these back to an outer eye, six to nine inches from the stem. The roots will establish themselves for the first year, and good growth will usually follow. The strength of a tree depends mainly on its roots. These must not be overtasked at first, or the tree will suffer seriously. Next year, late in July, cut back to the sixth leaf all shoots springing from the main branches which run inwards; keep the centre open, well exposed to the light, sun and air, and allow the main branches to develop themselves freely. In the winter cut all shoots not needed back to two or three eyes. If more boughs are needed, shorten the leading shoots, always cutting just above an outer eye. Make the tree as even as you can by shortening leading shoots on opposite sides. Never allow boughs to cross or to interfere with one another. If boughs are void of a fair proportion of shoots and spurs, they should be stopped. Be careful to admit the sun fully on the south side. Cut off all shoots springing from the central part or on the lower part of the branches of old standards. If young standard trees are well planted, carefully fed and pruned, the stems kept clear of weeds and grass, they can be brought into comparatively early bearing. Where irrigation is possible, let a stream of water that has flowed some distance over the ground be turned in dry weather on to their roots, or let liquid manure be given after rain; the effect will be surprising. But beware of very cold or stagnant water!

Early pears are probably the most profitable for orchard planting. The following are reliable:—

Six Market Orchard Standard Pears selected by Messrs Bunyard: Hessle, Fertility, Williams' Bon Chrétien, Beurré Capiaumont, Durondeau, Pitmaston Duchess.

Messrs Rivers' list of seven: Beacon, Bon Chrétien, Clapp's Favourite, Fertility,

Conference, Marie Louise d'Uccle, Vicar of Winkfield.

The list of an eminent firm in the south is as follows:—

Bon Chrétien, Hessle, Pitmaston Duchess, Louise Bonne of Jersey, Emile d'Heyst, Marie Louise. At the Pear Conference (R.H.S.), 1885, Bon Chrétien had 50 votes, Louise Bonne 46, B. Capiaumont 38, Hessle 30. Thus, William's B. C. has 4 votes, Hessle 3, Pitmaston 2, Fertility 2. Personally, I prefer Pitmaston as a bush, the fruit being so large. It is a pear for a good market, not a coster's fruit. Ten trees of three varieties would make a good orchard. Vicar of Winkfield or Verulam might be added for a later Stewing Pear. The bloom of Marie Louise is so tender that I prefer Marie Louise d'Uccle, a very good cropper; the fruit is sometimes sold as Marie Louise. The list of 1885 is hardly up to date. Louise Bonne does not do well with me as a standard, and I should substitute Fertility. Clapp's Favourite is also very promising.

If the plantation is of any size, do not put two trees of the same variety close together. Some varieties are self-sterile, yet quite capable of cross-fertilisation from the pollen of other varieties. Bees should be kept close at hand to fertilise the blooms.

The following is Mr Radcliffe Cook's list of orchard standards for Perry (see his "Cider and Perry"):—

Barland, Moorcroft, Red Pear, Taynton Squash, early varieties.

Langland, Yellow and Black Huffcup, midsummer.

Blakeney Red, Butt Pear, Oldfield, Pine Pear, Rock Pear, late.

It is said that in France there are more than 1500 varieties of Perry Pears. We must "wake up" and grow the best varieties.

PYRAMIDS

No one should plant high standards except under special circumstances; pyramids are a part of almost every large and good fruit-garden. In moist, strong soils they should be on the Quince Stock. In light soils the Pear Stock alone has a chance. Some trees succeed only as bushes, others can be trained as pyramids. The lists of the leading nurserymen usually refer to the habits of each tree. Buy trees trained as pyramids direct from the nursery. If you prefer maidens (trees one year old) train as follows: In early spring, after planting, stop the tree slightly, and encourage growth; next winter cut it down almost to the stock. A strong shoot from the base must now be made the leader and the central stem. Next winter cut this back to within 18 inches of the ground. The highest shoot next season must be trained upwards by a straight stake; the side shoots will form branches. These in September must be brought (by stakes) into a horizontal position. The stronger must be more depressed, the weaker may be left for another year. Bend into position before the sap sinks. In winter reduce side shoots on branches to two or three eyes. Cut the leading shoot 12 or 15 inches (according to growth or soil) above the branch below it, so as to produce fresh branches. Bend these down as before. As the tree progresses, the leading shoot may be stopped in summer when it has grown a foot, so as to throw out more branches; it may grow another foot upwards by September, and also send out fresh branches. Every care should be taken to keep an upright and straight stem. In summer pruning check the upper branches before the lower, stopping the terminal shoots so that they shall not spread out further than those below them. Stop them when they have grown 8 or 10 inches, removing the top. Any shoots from the branches (laterals) must be reduced to six or seven leaves about mid-June (on young trees), so as to open the tree and concentrate growth on necessary parts, and also to produce bloom-buds. These may form near the base. In winter reduce to two or three eyes.[4] Pyramids on the Pear Stock in strong soil reach a height of 15 to 25 feet, but such trees are hard to manage. Weak growing sorts might be tried. The larger trees would need annual root-pruning (half a side each year) to secure good crops. Train pyramids from the nursery in a similar way, keeping the upper branches in subjection to the lower, taking care to let light into every part of the tree by summer pruning. Pyramids on the Quince should be not less than 10 feet apart, 15 in strong soil with strong sorts (such as Pitmaston Duchess, or Duchesse d'Angoulême); on the Pear Stock in similar soil for strong sorts 20 feet apart. Avoid crowding. Lift or root-prune rather than crowd. Do not plant two trees of the same variety close together. The pollen of a different sort may make each tree more fruitful. Have hives of bees at no great distance to promote fertilisation.

The following are good sorts for pyramids:—

Citron des Carmes (on pear) early, Williams' Bon Chrétien, Clapp's Favourite, Marie Louise d'Uccle, Duchesse d'Angoulême, Durondeau, Fondante d'Automne, Beurré Hardy, Beurré Superfin, Maréchal de la Cour, Doyenné du Comice, Princess, Josephine de Malines, Beurré Rance.

Cooking Pears.—Bellissime d'Hiver, Vicar of Winkfield, Verulam.

Others might be added. Some of these also do well as bushes.

COLUMNAR TREES

are pyramids on a smaller scale, kept well in check by lifting or root-pruning, more like a column than a pyramid. In light soil this work would not be needed. They are adapted for small gardens, and, well managed, may be very useful. Plant from 8 to 10 feet apart.

ESPALIERS

in the open ground (according to some good growers) are the most economical of space, but I do not care much for them. Train at first from maidens as for a pyramid, keeping one upright shoot and guiding one branch each side in an almost horizontal position. Cut back the leader once a year at first at about 12 or 15 inches from the branch below to one bud just above the buds whence the branches are to spring. From this one bud the upright leader will grow. The branches should be about a foot apart. Stop the topmost in summer (if very strong) to divert the sap into other parts. Stop strong horizontals to strengthen the weak and to promote

[4] See elaborate account in the "Watson's G.'s Assistant," vol. iv. p. 116.

fruit-buds. Stop shoots on the branches late in June or in July at six full leaves, if the tree is flourishing, but not otherwise. Equalise the sap as far as possible. Espaliers may be bought from the nurseries, saving several years. Plant 15 or 20 feet apart according to ground and tree. Support with rails or stout firm stakes placed 2½ feet from the walk; place the tree 3 inches from the stake on the side of the path. Keep the trees low to prevent shade on the garden; 5 feet is high enough. Prune established trees in July; cutting back fresh shoots (laterals) to six leaves, and opening the tree and fruit to the sun, removing shoots not needed. Reduce to two or three buds in winter; with a small saw cut back large lumpy pieces the growth of years.

HORIZONTALS ON WALLS

should be trained as Espaliers. They are better for a low than a high wall. The branches should be about a foot (four bricks) apart. In some old gardens, enormous Horizontals may be seen with the branches at distant ends turned upwards. The lower branches are horizontal as far as the space allows, then turned upwards. This change checks the sap, lessens luxuriance, and promotes fruit-buds. But there often is excessive growth in the upper parts. These upper shoots must be pruned before the lower. Such trees are called Palmetto Verrier, and are scarcely to be recommended.

FAN-SHAPED TREES

are adapted to high walls. Tomatoes or other fruits may be grown below in the vacant spaces. By planting a standard against a high wall, it will soon be covered if fed and duly trained. Cut the tree back as an orchard standard after planting. Keep the boughs well away from each other, 12 inches or more apart. If a wall is shaded with foliage it derives little heat from the sun. Stop the gross upright shoots early in the season to spread the sap, and summer prune in July. Keep the branches close to the wall, and complete pruning in winter. These trees must be on the Pear Stock. The choicest sorts, such as Doyenné du Comice, Beurré Superfin or Diel should be selected for a south wall. Prune the upper parts before the lower. Wires may be placed on the walls 1½ inches out, with an interval of 12 inches or more between each wire.

BUSHES

are of great value, either in a plantation or a garden. In good soil, even those on the quince grow large, and may need root-pruning or moving. In poor soil, with gravel or chalk not far below, bushes on Pear Stock must be moved every few years, and well fed. Rotten manure given in the autumn will attract and feed the roots. Fruit on low bushes is less affected by strong winds. Some sorts do better as bushes than as pyramids; bushes, too, are more under control. A maiden tree after planting should be allowed to grow for a year unchecked, to establish the roots. In winter cut the tree back to within a foot of the ground. In the spring it will throw out vigorous shoots. Select three or four of these, and fix them in position with stakes, removing the others. Next winter cut these back to an outer eye, leaving six or nine inches of each branch from the stem. Other branches will soon follow. Time will be saved by buying bushes from the nursery. Keep these as open as possible, especially on the south side and the centre. Each branch should be a foot apart. Summer prune in July and winter as before. Stop the branches in summer, if growing rapidly, to produce fruit spurs, and in winter cut back to strong wood (to an outer eye). All new wood will thus be feathered during the following year. Some bushes are very diffuse and need much room, *e.g.* Catillac and Uvedale St Germain. Bushes on quince should be eight to twelve feet apart; strong growers, such as Pitmaston, Duchesse d'Angoulême, Catillac, should be even more in good soil, if root-pruning is not to be practised. The following are good as bush trees:—

Bergamotte Esperen on Wall

supply of some kind at hand. Not a drop from his stables, etc., should be wasted in summer. In a drought it may save his trees.

But rank or fruitless trees of any age, as a rule, need no manure. If there is a heavy crop, feed well when the growing season is over. Pears are gross and thirsty feeders. Messrs Rivers[5] recommend "that a peck of soot should be strewed on the surface in a circle 3 feet in diameter round each (dwarf) tree in March. Pears on the Quince in a light, dry soil should have the surface round the tree covered during June, July and August, with short litter or manure, and in dry weather be drenched once a week with guano water (1 lb. to 10 gallons), and equal parts of soot, which must be well stirred before it is used. Each tree should have 10 gallons poured gradually into the soil. Lime rubbish or chalk should be added wherever there is any deficiency." If it be possible, in dry weather allow a stream of water to flow by their roots, or in any case give liquid manure. The roots should never be dry; cracking often follows rain just after a drought if the roots are dry. Soot is a safeguard against insects, and is supposed to give colour. Dr Griffiths (in "Special Manures for Garden Crops," p. 101) says: "Nitrogenous manures are requisite for backward, potash and phosphates for forward trees; the former aids growth, the latter develops bloom, the sugar in the fruit, and the ripening of the wood. Pear trees are aided by a manure containing four parts (by weight) of kainit and one part of superphosphate—4 lbs. of this mixture to be given in the spring to each tree after pruning. If the trees are backward, water once a week with a solution containing 1 oz. of nitrate of soda to 2 gallons of water." If basic slag and kainit are given, autumn is the time, as their action is slow. Nitrate of soda is good on hot, dry, and chalky soils.

PEARS FOR A PRIVATE GARDEN

If the space is small, try cordons or bushes. If three are enough, Fertility, Pitmaston, Josephine de Malines; if six, add Durondeau, Bon Chrétien, Comice; if nine, add B. Hardy, B. Superfin, Verulam; if twelve, B. d'Amanlis, Louise Bonne, B. Clairgeau; if fifteen, Jargonelle, Clapp's Favourite, B. Diel; if twenty, Doyenné Boussoch, Marie Louise d'Uccle, Maréchal de la Cour; if twenty-three, Glou Morceau, Winter Nelis, Passe Crassanne; if twenty-six, Comte de Lamy, Dana's Hovey, Thompson's; if thirty, Doyenné d'Été, Emile d'Heyst, Baronne de Mello, Easter Beurré or Olivier de Serres.

EXHIBITION PEARS

Size is of importance as well as perfection in every point. Coarse pears of inferior quality rarely win. Choice must depend on the time of year when you compete. The same fruits cannot be sent to several shows; they are certain to be bruised and to suffer in some way. The following are the chief pears for exhibition:—

August and September.

Beacon. | Souvenir du Congrès.

[5] See "Miniature Fruit Garden," p. 64.

Flemish Beauty.	Clapp's Favourite.
Bon Chrétien.	Marguerite Marillat.

September and October.

B. d'Amanlis.	Bonne d'Ezée or Brockworth Park.
Beurré de l'Assomption.	Triomphe de Vienne.

October.

B. Hardy.	Marie Louise d'Uccle.
D. Boussoch.	B. Superfin.
Louise Bonne.	

October and November.

Beurré Alexandre Lucas.	Maréchal de la Cour.
Emile d'Heyst.	D. du Comice.
B. Diel.	Pitmaston D.
Beurré Fouqueray.	Magnate.
Duchesse d'Angoulême.	Conference.
Durondeau.	Marie Louise.

November and December.

Thompson's.	B. Sterkmans.
Nouveau Poiteau.	B. d'Anjou.
Princess.	Glou Morceau.
Fondante de Thirriott.	General Todleben.
B. Baltet Père.	

January, etc.

Nouvelle Fulvie.	Passe Crassanne.
Bergamotte Esperen.	President Barabé.
Olivier de Serres.	Easter Beurré.
B. Rance.	

COOKING PEARS FOR EXHIBITION

December and April.

Uvedale's St Germain.	Bellissime d'Hiver.
Catillac.	Directeur Alphand.
Verulam.	

Size is the chief point in cooking pears, then equality of excellence. Size is produced by careful culture and good feeding in good soil.

The dates above are only approximate.

Pear—Maréchal de la Cour

Pear—Marguerite Marillat

PEARS FOR APPEARANCE

Doyenné Boussoch is perhaps the most handsome of all pears, but does not last long. Marguerite Marillat (September) is large and handsome, so are B. Clairgeau, B. Sterkmans, B. Mortillet, Souvenir du Congrès, B. Baltet Père (very turbinate), B. Giffard, B. Hardy, Louise Bonne, and others.

PEARS FOR QUALITY

Much depends on the season, soil and situation. In a cold season, even pears of good quality are only fit for cooking. Thus used, they are often excellent. The sweetest of all pears is Comte de Lamy. Dana's Hovey (of American origin) is perhaps its equal. D. du Comice, B. Hardy, Marie Louise, Josephine de Malines, Winter Nelis, Bon Chrétien, B. Superfin, Thompson's, Fondante d'Automne, are among the best. A warm autumn makes a vast difference. B. Diel then becomes first rate, so do Passe Crassanne, Olivier de Serres, Bergamotte Esperen, B. d'Anjou, B. Sterkmans, and others.

COOKING PEARS

Growers should keep in mind that dessert pears often cook well if gathered before they are ripe. Stewed pears are excellent food in every way; pears that do not ripen well can be utilised thus. There are special sorts pre-eminently good. Verulam and Bellissime d'Hiver, very fertile as bushes or cordons, keep and cook well. Catillac and Uvedale's St Germain are very large, the latter often enormous; the fruit sometimes exceeds 2 lb. if the tree is well fed. The two last are spreading as bushes, but do well as cordons. Bellissime d'Hiver was the favourite C. pear of the famous Dr Hogg. Vicar of Winkfield is also good, but not so lasting. Cooking pears should begin in September and last until April. B. Clairgeau is regarded by the R.H.S. as a cooking pear. It is free-bearing and handsome, but not lasting. Directeur Alphand (new) is described as very large, but needs sun to ripen.

EARLY PEARS

These are not important (except for sale), as so many fruits of other kinds are usually abundant. Doyenné d'Été is the first in. Double-grafted on the Quince, it is very fertile. Next comes Citron des Carmes, a great French favourite. The fruit of this is said to be fine when the tree is double-grafted. Crawford, a favourite Scotch pear, is regarded as its superior north of the Tweed. Jargonelle is also a Scotch favourite, especially in Perth, where every vacant wall space is said to be soon occupied by this pear. It is grown, too, as a standard on the free stock, but does not love the Quince. If double-grafted, the leading shoot pinched as well as the side shoots two or three times in the season, it will bear well. Beacon and B. Giffard are also August pears. Later on come Clapp's Favourite, Bon Chrétien, and many others. Early sorts should be gathered before they are ripe. Mr G. Bunyard recommends that early pears as well as early apples should be laid in heaps, covered with nettles or straw, and "sweated," to improve their appearance. They are said to colour well treated thus.

LATE PEARS

Are often worthless until they are in the kitchen; yet a warm autumn makes some of them delicious. The best of all is Josephine de Malines. The tree does well as a standard or bush, and the demand for the fruit is sometimes great. With care it will last to March. Next comes Winter Nelis, not so hardy; then follow Nouvelle Fulvie, Madame Millett, Passe Crassanne, Olivier de Serres, Easter Beurré, and B. Rance. A new sort, President Barabé, has received a First Class Certificate from the R.H.S. Late varieties must be allowed on the trees as long as possible, and be *well* protected from birds. Great care must be taken in handling and storing. Bruised pears soon rot.

PEARS FOR COTTAGERS AND SMALL FARMERS

The following were selected in 1892 by the R.H.S. on the advice of forty experts: *for eating*, Jargonelle, Bon Chrétien, B. d'Amanlis, Louise Bonne, Durondeau, Marie Louise, D. du Comice, Pitmaston Duchess; *for cooking*, B. Clairgeau, Catillac, Uvedale's St Germain, Verulam. But Marie Louise is a poor and uncertain bearer.

SYNONYMS

When fruit trees have numerous names, they certainly are popular, probably good.

Passe Colmar has twenty-eight, chiefly French; grown in a rich warm soil it is a first-rate dessert pear (November). The tree is vigorous and makes a good pyramid.

B. Diel has thirteen: among the French it is Beurré Magnifique. It requires a good season here.

Uvedale's St Germain (Belle Angevine of the French) has twenty-two, chiefly French. Yet it was raised in 1690 by Dr Uvedale, a Schoolmaster of Eltham in Kent.

Windsor, a very old English pear, mentioned in 1629, yet of French origin, has eleven. The fruit is large and greenish-yellow, flushed, but soon becomes dry and worthless. In good soil it grows and bears well (August).

White Doyenné has fourteen, a fairly good September and October pear, rather large, a good bearer, "flesh white, but somewhat acrid and gritty" (Barron).

Vicar of Winkfield has twelve. A long large fruit often twisted, fairly good for baking, from November to January, "second rate" (Barron).

B. Rance has six. A long, largish, late pear, sometimes very good.

Wardens, a name given to pears which never melt, are long keeping, and used for cooking only. The name comes from the Cistercian Abbey of Warden in Beds. Parkinson's Warden is now Black Worcester. There are Spanish, White and Red Wardens.

Bishop's Thumb was originally called Bishop's Tongue, It was a favourite in 1690, and is still a favourite. The tree is hardy and a good bearer, the fruit long, firm, melting, sweet (October, November).

Brown Beurré has ten; an old favourite, which requires a wall or very warm site (October).

Chaumontel has nine, requires a very warm climate. Better in Jersey than in Britain.

Pear—Beurré Diel

Easter Beurré has twenty-two, most of them French. Good if grown in good soil and in a good season. It does not grow well on the Quince.

Flemish Beauty has seventeen. The fruit is large and sometimes russetty and flushed crimson; good only when gathered before it is ripe (September and October).

Louise Bonne has seven. Raised at Avranches in Normandy (1788), it curiously is called L. B. of Jersey.

Maréchal de la Cour has six, large and good. "One of the finest" (Dr Hogg).

Napoleon has fourteen. "Second rate" (Barron).

Red Doyenné has eleven, chiefly French. The fruit is superior to White D. (November).

Glou Morceau has twelve or thirteen, chiefly French. It is excellent in a warm soil and site (November and December).

PEARS FOR PERRY

Our people are beginning to discover that we can and ought to make as good Cider and Perry as is made in any country. Mr Radclyffe Cooke in his "Cider and

Perry" gives the following list:—

Early Varieties.

Barland.	Moorcroft.
Red Pear.	Taynton Squash.

Midsummer.

Langland.	Yellow and Black Huffcap.

Late.

Blakeney Red.	Butt Pear.
Oldfield.	Pine Pear.
Rock Pear.	

Sixty varieties appear in the List sent to the Pear Conference of the R.H.S., October 1885.

GATHERING AND STORING

Mid-Season and late pears should be gathered in dry weather as soon as they come easily from the tree. Lift gently, and gather by degrees as the fruits ripen, those on south side first. Use padded baskets, and treat good fruits with loving care. Beware of piling a large quantity in one basket, of turning or rolling out instead of handling by the stems. With high pyramids Heathman's combined ladder-steps may be needed. Pears should be put away quite dry in a dark and dry place, where the temperature is as even as outside wooden or other walls, and thatch above can make it. Perfect and fine fruit should be wrapped in tissue or other paper and placed singly on shelves or in shallow drawers or boxes. Boxes are excellent for late fruit. For storing they should be only deep enough to hold one layer of fruit. Scott recommends clean bran, others dry silver sand, to put among the fruit so as to absorb any moisture. The ripening may be hastened by placing the fruit in a gently warmed room, or on hot water pipes in a greenhouse. "Sorts dry and tough carefully ripened in warm drawers or on the shelves of a warm cupboard become deliciously melting and rich. A heat from 60° to 70° is about the proper temperature" (Scott). Fruit pecked, bruised, or injured in any way should be kept apart and got rid of without storing. White tissue paper,[6] glazed on one side, the fruit resting on the glazed side with another sheet on the top, the glazed side downwards, is useful where a large amount of fruit is stored on shelves or trays. Orr's Patent Trays, sold by John P. White, Bedford, are excellent for storing. The trays fit on each other, and single trays are readily moved, so that the fruit on each tray can be examined without being handled.

PROTECTION OF FRUIT

As trees must be protected against hares and rabbits, so must fruit be from other enemies. Birds in some seasons are most destructive, attacking the finest fruit, pecking a piece out near the stalk. Such fruit soon decays. Wasps and blue-bottle

[6] See Watson, vol. v., "Storing."

flies feast on ripe or injured fruit. Mr Cheal in his "Fruit Culture" recommends that galvanised wire netting be put over the whole ground. This may do for small plantations, not for large, nor for places where the trees rise beyond 7 feet. Many use the Cloister Fruit Protector of perforated celluloid. This protects peaches, apples, pears, etc., from birds, wasps and snails, but the cost is heavy. Muslin bags kept carefully from year to year are good. The fruit rests in them and grows. Nets made in different sizes might be put over bush trees on stakes. They last if kept dry. The gardener, too, should have a gun and use it at dawn and daily. Messrs Bunyard recommend a trap like a lobster pot made by Gilbertson & Page, Hertford, to be baited with soaked bread. This trap takes birds alive. The house-sparrow and the bullfinch are the chief, but not the only, enemies. Robins, hedge-sparrows,[7] etc., might be released. Cut ivy carefully back, and encourage winter nets and sparrow clubs. Frost is another foe. Cordons might be protected by hoops covered with tiffany, Russian canvas, mats, or netting; bushes by nets, mats, etc. A movable coping over a wall is often useful. But if strong colonies of bees are close at hand, they will rarely fail to fertilise some blossoms. In fine intervals bees come out in crowds, and do great good. Queen wasps and wasps' nests should be sought and destroyed. Country children will find them for a small reward.

WINTER AND SPRING WASHES

If the fruit-blossoms survive frost, cold winds and rain, enemies of a different kind await them. It is necessary to spray or wash the trees if these enemies are to be kept at bay.

1. The following mixture is recommended by the Board of Agriculture: "To prepare caustic alkali wash, first dissolve 1 lb. of commercial caustic soda in water, then 1 lb. of crude potash (potashes or pearl ash of oilmen) in water. When both have been dissolved, mix the two well together, then add ¾ lb. of soft soap or agricultural treacle, stir well, and add sufficient water to make up 10 gallons." As the wash has a burning effect on the hands, the sprayer should wear gloves and be careful. The Eclair hand-spraying pump, supplied by Clark & Co., 20 Great St Helens, E. C., sends a spray like a mist. The cost is about 35s. We have used it for years, and the same firm repairs it well. This mixture with us, though easily sprayed, has not been a great success. If used, it should be applied in February, just before the buds open.

2. The Bordeaux Mixture is used for spraying by some, and is recommended by Messrs Bunyard. It is a good fungicide as well as insect-enemy. The following is the receipt: Sulphate of copper 6 lbs., unslaked lime 4 lbs., water 50 gallons.

Dissolve the sulphate of copper in a wooden vessel, pouring in sufficient water to cover the coarse bag in which the sulphate should have been placed. Attach the bag by means of a string to a rod placed across the vessel, and let it hang in the water. In another vessel add water gradually to the lime until a thick paste is formed; when cool mix the two together in a third vessel, and add water up to 40, 50 or 60 gallons. If the mixture is properly made, a clean knife blade held for one

[7] Hedge-sparrow smaller, duller in colour, eggs bluish green, builds in hedges; house-sparrow, eggs white, with brown spots, nests in trees and buildings.

minute in the solution should remain unchanged; if coated with copper, add more lime until no copper adheres to the blade. Stir the mixture constantly while spraying and use it fresh. Spray the trees when the buds are first expanding. Messrs Bunyard (Fruit Catalogue, 1901-2) recommend "6 lbs. of pure sulphate of copper, 4 lbs. fresh unslaked lime, and 22 gallons of water, the sulphate to be put in a piece of sacking or light cloth, and hung by a string from the top of a barrel containing 18 gallons of water, a few inches below the surface so as to dissolve. Then slack 4 lbs. of fresh lime in as small a quantity of water as possible, the water being added very slowly, until slaking is completed; then slowly make up to 4 gallons. When cool, thoroughly stir and strain slowly the milk of lime into the copper solution, stirring well while mixing for another minute or two; it is then fit for use as a winter spray. It should be used when freshly made, (*a*) Apply before buds start to all fruit trees with the 22 gallons mixture. This can be diluted to a 30, 50 or 60 gallons mixture for spring or summer use. (*b*) Spray again just as the petals drop with the 60 gallons mixture. If made and applied as above (within ten or twenty hours) it adheres closely to the wood and foliage; treacle need not be added." This adhesion is of vast importance, as lime is abhorred by stem-borers (*e.g.*, the goat and leopard moths) as well as by all insects. The double application of lime is also helpful. In the United States Paris Green is sometimes added, and is no doubt useful; the proportion must be very small.

3. For many years I have painted my trees in winter with the following mixture: one bushel of lime, half a bushel of soot, a quart of paraffin, a pail of cow dung, a pail of clay; melted grease is sometimes added, and the whole worked into a paint and then put on the trees. Treacle might be substituted for the cow dung and grease. This has proved a valuable preventive. The lime and soot gradually falling off, leave the bark clean, and enrich the soil below. But painting is a much longer process than spraying with (1) or (2). Apples have subsequently been sprayed with Paris Green, and pears might also be.

INSECT ENEMIES

1. The pear oyster scale is very injurious, especially on walls, if not checked at an early stage. The covering of the female is like a small oyster scale, hence the name. Scrape off any rough bark in winter, and apply the alkali or one of the other washes as a preventive. In May and June affected parts might be brushed with ¼ lb. of soft soap in a gallon of water. Tobacco or lime water might also be applied. Paraffin largely diluted may be used, but is dangerous in excess. Messrs Rivers in "The Miniature Fruit Garden" (p. 144) say: "Washing the parts affected with a mixture of soot, lime and sulphur will remove the roughness and restore the tree to health; the above mixed with skim milk is more enduring." As a believer from experience in soot and lime, I prefer this receipt, if the trees were not washed in winter.

2. The Blister Moth makes brown blisters on the leaves. It may be kept from laying eggs on the tree by syringing occasionally with soap-suds. Spraying with Paris Green just after the fruit is formed will do good. Half an ounce of best paste to 10 or 12 gallons of water, with some fresh lime added, will suffice for small

gardens. Spray only in fine weather just after the petals have fallen. Paris Green is arsenic, and may poison bees if used too soon. The sprayer should avoid breathing over the mixture when making it up, should use gloves, work from windward, and not allow any spray to reach his flesh. A second spraying for this and other insects is often useful. Blundell, Spence & Co. (Ltd.), Hull, supply good paste. Price ½ lb. 1s., less for larger quantities. See also No. 3.

3. The Pear Leaf Mite causes small blisters on the leaves, but not the tunnels or galleries of the Blister Moth. It winters in the bud scales, and emerges in the spring. If the trees are washed and syringed, the attacks will be lessened. In (2) and (3) collect the blistered leaves as soon as seen, burn them and spray or syringe at once.

Miss Ormerod recommends a dilute paraffin emulsion sprayed over infested leaves. Dissolve ¼ lb. of soft soap in a gallon of water, add this while boiling to two gallons of paraffin, churn the whole with syringe or small pump for ten or fifteen minutes to make a perfect mixture. For spraying add 12 gallons of water to each gallon of the emulsion. Stir well while spraying, and try the mixture on a branch or two lest it be too strong; if so, add more water. This emulsion is good for the Blister Moth and the Slug-worm.

4. The Slug-worm is so called from the similarity of the larva of this sawfly to a small black slug. The worms feed on the upper surface of the leaves. Dust with quick lime two or three days in succession, or syringe with strong soap-suds and some tobacco water. Clean with pure water in a few days. The paraffin emulsion (No. 3) might also be used. Quick-lime scattered around the roots and forked three or four inches into the soil may destroy their cocoons. But beware of excess. The remedy may be worse than the disease.

Insects that attack leaves will also eat the skin of the young fruits if conveniently placed for them.

5. The Pear Sucker is a jumping plant-louse which early in the season sucks the juices of the tree about the axils of the leaves. They are covered with the exudations of the sap, which often drops on the ground. The visits of the ants should call attention to this pest. Syringe well with soft soap and water, ½ lb. to 4 gallons, and add tobacco water. Remove all rough bark (their hiding-places) in winter.

6. The Pear Gnat Midge (*Diplosis pyrivora*) may readily ruin a crop if unchecked. It is a recent importation among us. Both here and in the United States it is spreading with alarming rapidity. It is a small two-winged fly, with a black body having lines of yellow hair. The female pierces the flower-buds and lays her eggs in them. These soon hatch, and the young tiny grubs eat their way into the embryo fruit, keeping to the fleshy part, leaving the core and seeds alone. The pears turn brown, and then black. Cut them open, you will notice maggots. The fruit bursts or falls, the maggots form silken cocoons in the soil in which they pupate, and remain till the blossoms begin to expand next spring. Mr J. Fraser (editor of *Gardening World*) has kindly sent these details, and recommends (1) that the injured fruit be gathered and burnt; (2) that two inches of the ground beneath the trees should be taken up and burnt; (3) that kainit should be distributed round the

trees in autumn. Kainit is said to keep off wireworm, and is recommended in the United States as a preventive against this pest. I think the mixture No. 2 or No. 3 should also be used, as insects may be deterred by the scent. Lime and soot spread over the ground in winter would probably do good.

7. Weevils devour leaves, buds, young shoots, even the skin of fruit. They feed by night, and may be shaken into a cloth off bushes. Lime and soot may lessen their attacks, either as a wash No. 2 or 3, or spread lightly round the stems, or as a powder over the leaves.

A special bellows for distributing any dry powder (as sulphur, lime, soot, etc.) can be had from De Luzy Fréres, 44A Harold Street, Camberwell. The price is 7s. 6d., carriage paid.

As a general rule insecticides should be applied in the evening or after the sun is down. Early and late visits to the trees are best for finding them feeding.

8. *Wasps*, after a dry spring, may be very numerous. Their nests often hold many thousands. Large numbers may be destroyed thus: place a hand-light upon bricks, make a small hole in the top of this, and over it put a sound and closely-fitting one. Fruit cut open should be thrown beneath the lower light. The wasps often go up through the hole, and do not return. Their buzzing attracts others. Destroy by burning sulphur beneath, or by drowning. A glass destroyer on a similar principle is sold in china-shops. Open-mouthed bottles filled with beer sweetened or water sweetened with treacle will lure many to destruction. Queen wasps in spring and wasp-nests must be noticed and destroyed. Fasten a piece of cloth soaked in a solution of cyanide of potassium (a small quantity dissolved in hot water), and put it in the nest; all the wasps will be killed. Dig out the grubs. This is a deadly poison, and should be handled only by an expert. The emanation from the solution must not be breathed. Tar does almost as well. A nest may be partly dug and flooded at night. A clean wine bottle (half-filled with water) inserted in the place of the nest (the top of the neck level with the surface of the ground) will probably capture all stragglers. Some make a heap of injured fruit and syringe the wasps with nicotine soap, eight ounces to a gallon of hot or cold water. This plan kills quickly, but the fruit no longer attracts. Squibs a half-inch in diameter, three inches long, made of gunpowder moistened with water, one-fourth of flowers of sulphur added, mixed into a paste, wrapped in brown paper, and tied at one end, are good for the work. After dark, light the squib, push the lighted end into the hole, put a sod over, and ram it in to confine the fumes. In a few minutes dig up and destroy the grubs, then fill up the hole. If the nest is high up, attach the squib to a stick, light, and keep it close (while burning) to the entrance. Young gardeners enjoy this squibbing process.

THINNING FRUIT

If you wish for fine fruit or a crop every year, trees must not be overworked, especially in their earlier days. Thin whenever there is a large crop, but do not begin too soon, as some fruits are not fully fertilised, and may fall. Never let fruits touch each other. As the fruits mature, give any grub-eaten to the pigs, and use inferior pears for cooking purposes. Grub-eaten fruit must not lie on the ground.

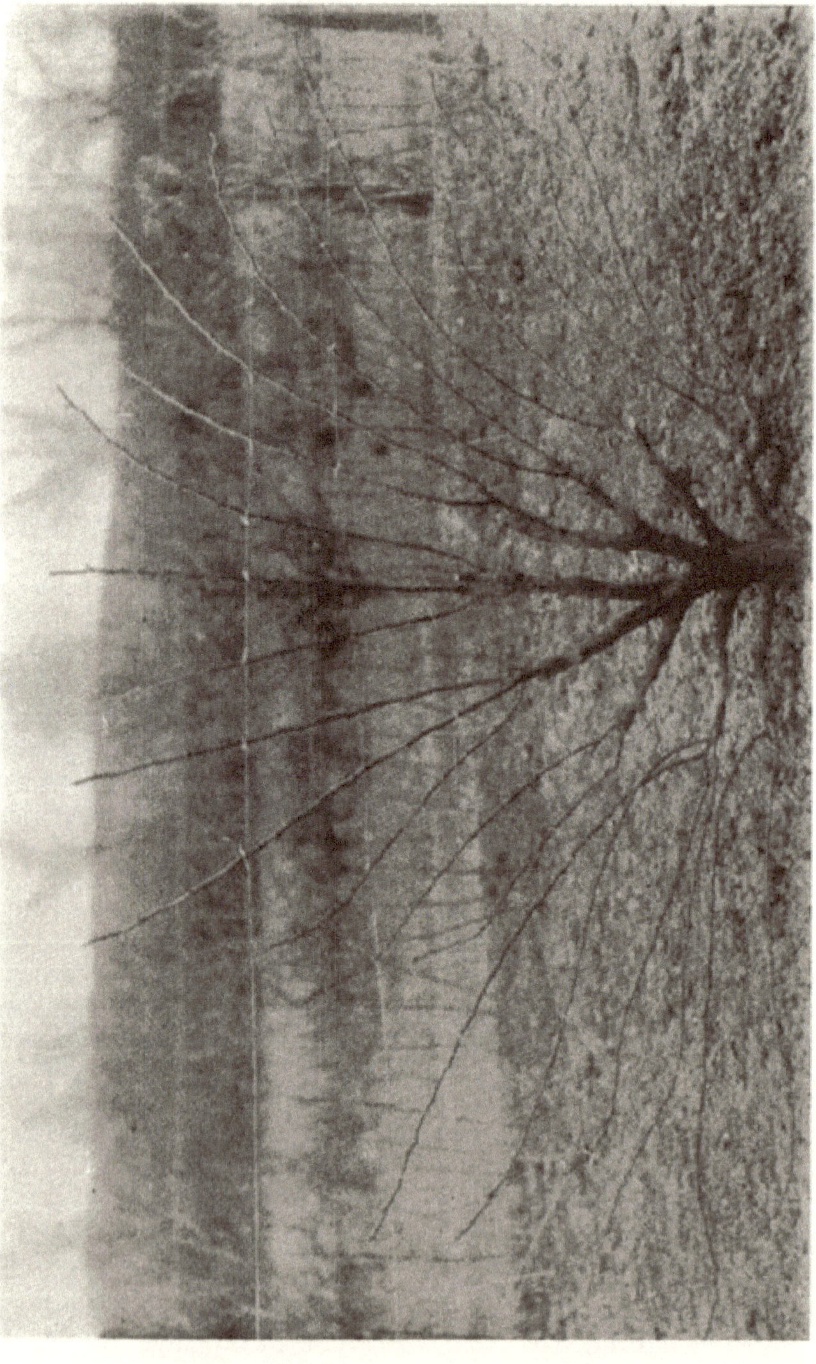

Fan-shaped Pear Tree, One Year After Grafting, showing the Length of Resulting Shoots

SUMMER, WINTER, BRANCH AND ROOT PRUNING; LIFTING

Summer pruning rests chiefly on the principle that the trees should always be open more or less while in leaf to the sun, the light, and the air. So cut out *at any time* branches that crowd the tree or threaten to cross other boughs. Cut from below, so as not to tear bark away. Pears do not bleed from being cut. In July, when the growing time is almost over, cut back to six or seven leaves any strong shoots springing from a main branch, or in cordons, from the stem. If they shoot again, they should again be stopped. In late autumn or winter look over the trees, reduce the shoots to two or three eyes, taking care not to remove bloom buds. Early in the summer, and at any time, remove from the trunk and boughs any shoots threatening to crowd or shade the centre. Keep the tree (especially the centre) open to sun and light. Even large standards are improved by summer pruning. Tree-pruners should be used where the shoots are out of hand-reach. Root pruning is also essential in strong soils where trees are too rank in growth and produce wood rather than fruit. Trees of all kinds may be root pruned with advantage in such soils, and also where the lower soil is bad. Open a trench 20, 30 or 40 inches from the stem (according to size of tree) until the coarse roots are reached. One-third the distance from the stem that the trees are in height is a rule suggested by a recent writer.[8] Cut back such roots with a sharp knife; drive the spade under the stem (if possible) to cut the tap roots, and any others going downwards. Open a trench half round one year, and if necessary attack the other half next year. Be careful not to prune too hard at first, or to injure the fibres. Begin in mid-October. If the ground below is very dry, give warm or rain water. Fruit blooms will probably appear next autumn. If young trees grow very luxuriantly, they may be lifted at the end of October with advantage. Cut the tap root and replant at once. Exposure of the roots is dangerous to vitality. Persons who prune their trees only in winter usually grow wood rather than fruit.

MARKETING AND PACKING

Marketing depends greatly on the neighbourhood. Colour, size and quality ensure a sale everywhere, but only a constant supply of good fruit will attract retail dealers or the London salesmen. Poor stuff will not sell at a good market. The early fruits may be sent in flats (with tops) lent by the salesmen. But these are often lost and involve trouble and expense. Non-returnable boxes to contain half a bushel or a bushel are now in use, but such boxes are too large for the better fruits. Californian pears come to us in good condition in boxes containing each a few dozen fruits, each fruit being separately packed in tissue paper. French pears are also sent in boxes evenly graded and packed in one, two, or three layers. Small boxes bought by the gross are not dear. The following list is taken from Watson, vol. v. p. 369.

[8] John Wright, "Profitable Fruit Growing."

Length.	Width.	Depth.	Gross	
			s.	d.
11½ in.	10½ in.	7½ in.	32	6
15 in.	6 in.	7½ in.	31	6
15 in.	11 in.	7 in.	50	0
15 in.	13 in.	4 in.	53	6
16 in.	8 in.	4 in.	28	6

—BLACKNELL.

In the larger boxes, strong paper should be put round inside to prevent bruising. All fruit, however sent, should be even in size, of good quality, not diseased or bruised. Pears are more attractive when well packed than apples. Placed with their heads against the two opposite sides in two rows with the stems toward each in a box of suitable size, they may be made to fit closely so as to travel safely. The better and later sorts should be bedded in wood-wool and wrapped in tissue paper, white or coloured, with a sheet of paper between each layer, and the whole firmly packed. Loose fruit are sure to suffer. The contents of each box must be made so firm as not to be moved in the slightest degree. The G.E. and other railway companies provide cheap boxes of a suitable size and allow similar boxes also to be used if nailed. They must not be corded. Wire hinges and a fastening in front have been suggested. Nos. 3, 4 and 5 (G.E.R.), 2s. 6d., 3s., and 4s. per dozen are the best sizes. They will hold 18 to 24 fruits. On G.E.R. 20 lbs. can be sent for 4d. to London; 1d. extra is charged for every additional 5 lbs.; delivery is included. Such boxes could be readily stamped with the grower's name. The companies assist growers by publishing the names of those who have produce to sell.

PEARS IN AN UNHEATED ORCHARD HOUSE

With skill and care pears may be successfully grown in an unheated orchard house. They may have apples for their companions, but not cherries, peaches, plums or apricots. The most convenient house is a span-roof from 20 to 24 feet wide, 10 to 12 feet high to the ridge of the roof, and 4½ to 6 feet at the sides. Ventilators should run round the sides 18 inches wide, and hinged at bottom; the top ventilators should be 3 feet wide by 15 inches, 7½ feet apart, on alternate sides of the ridge (Mr T. Somers Rivers, in *Royal Horticultural Journal*, vol. xxv., parts i., ii.). A good length for this breadth is 50 to 60 feet. A half-inch wire protection over the ventilators and an inner wired door may be as necessary (as a protection against birds), as it is for cherries. There should be a path made hard with clay and gravel through the centre. Some advise a concrete floor; others prefer to plunge their pots inside as well as out. A lean-to house from 6 to 9 feet wide against a south wall may be of great service. Cordons can be grown on the wall, or planted outside and trained indoors, like vines, near the glass. Trees in pots can also be placed there. With either house, some ground to which the trees in pots can be removed when all danger from frost is over is required. It should be warm and well sheltered. Maiden plants may be put into 8 or 10-inch pots in September, and cut back later on, but time is saved by purchasing older trees of nurserymen; 15 to 18-inch pots will be needed in a few years. If there is a concrete floor, the pots must be raised on bricks, that surplus water may pass off. If the pots are plunged, care must be taken that the water can run away. In June take them into the open air, plunge them in the ground within three inches of the rim, to keep them warm and moist, and to protect the trees from the wind. After the fruit is gathered, the trees should as a rule be repotted. Prepare a fresh pot with broken flints, etc., at the bottom, place a piece of turf on them, next a handful of soot, and some fine soil on that. Have ready some new soil made chiefly of good turfy loam, to which old mortar rubbish or road scrapings, wood ashes, guano, and bone-dust have previously been added. The whole should be well mixed. Then take the tree out with a ball of earth, remove the soil all round the ball with a pointed stick, shorten the rootlets around, and cut any coarse roots away with sharp pruning scissors. Place the topmost roots an inch and a half below the rim, then shake this compost among the roots, finally ramming the soil hard down into the pot. In two or three days soak the ball with rain or warm water. The trees are better in the house until re-established. Sprinkle the leaves daily with soft water. Close and keep the house moist. The pots can then be taken out and plunged once more. The house will probably be wanted. They must be carefully protected in severe weather; place ashes, earth, or manure around them. Another plan is to lay the pots on the ground and cover them with mats. Take them back to the house before the buds begin to move. Shape the trees in winter, and summer prune as may be necessary. They require syringing as well as rich feeding when carrying a crop. A mixture of poultry droppings or night soil (half a barrowful) added to the same amount of sifted soil and of wood ashes, with a peck of soot and a peck of bone dust, all made into a compost a few days before use, is a strong surface-dressing. A layer half an inch thick when the fruit is swelling should be given two or three times, and be watered down with a fine rose. Messrs Bunyard recommend cow manure mixed with malt combings, and (as an artificial) sulphate of ammonia.

Liquid manure (not strong nor cold) must also be given two or three times a week. The fruit must be thinned, and the trees never over-cropped. Large trees in 16 or 18-inch pots need the annual renewal of the soil rather than repotting. The flowers should be fertilised by the admission of bees, by shaking the trees in fine weather about mid-day, or by passing a light brush gently over the blooms from flower to flower. Change of diet as well as air, and frequent syringing with clear water (say Messrs Bunyard) are very necessary ("Modern Fruit Culture," p. 23). But a dry atmosphere is best when pear and plum trees are in flower. Syringing in the open air is good for all trees in dry weather after the fruit has set. The following is a good wash to be applied when the trees are brought into the house in January or February. Put a peck of fresh soot into a coarse sack, and hang it in a tub containing 30 or 40 gallons of water; leave it there for eight or ten days; then remove it and throw in half a peck of fresh lime. Mix well, then take off the surface scum. A decoction of quassia made by boiling 2 or 3 ozs. of chips to a gallon of water for twenty-five or thirty minutes (or steeped in soft water for twenty-four hours) added to the above is a useful insecticide. Syringe with this before the buds appear, but not again until the fruit is set, then once a week, or oftener, as occasion may require.

Espalier Trained Tree Cut Back for Grafting—the Grafts Inserted and Clayed Over

N. B.—Never repot until you have learnt that the ball and roots of the tree are thoroughly moist. Soak the ball, if necessary, for twenty minutes. In surface-dressing leave a space near the tree open, that you may see what water is wanted. Never give *strong* liquid manure. As severe frosts and dull weather sometimes occur in March when the trees are in bloom, some hot-water pipes (two rows of 4-inch) may be added if means allow. A span-roof house should run north and south. Only the choicest sorts should be deemed worthy of a house, such as Bon Chrétien, Souvenir du Congrès, B. Brown, B. Superfin, Louise Bonne, B. Hardy, Maréchal de la Cour, Marie Louise, D. du Comice, Josephine de Malines, Winter Nelis, Passe Crassanne, Bergamotte Esperen, and others.

OLD STANDARDS

Old Standards that have ceased to produce good fruit should be cut down to within a few feet of the stem. The young wood will soon bear better quality. The trunk should be well cleaned and washed.

IRRIGATION

Wherever possible, irrigation should be applied in dry weather. An aero-motor pump or engine of some kind may raise the water to a tank. It should be allowed to run over the ground for some distance to be warmed and aerated. Apply in strong soil only when the growing season is over.

LABELS

Labels add greatly to the interest and pleasure of a garden. Acme labels are popular. Those sent out by John Smith, Label Factory, Stratford-on-Avon, are also good. They may be attached by his copper wire, but those of the form of the rose labels with the name affixed at the top of a long spike are less likely to be lost.

AMERICAN PEARS

The chief pear in the States is the Bartlett, corresponding with our Bon Chrétien. A schoolmaster named Wheeler, of Aldermaston (Berks), raised it about 1770. A nurseryman named Williams brought it out. In 1799 one Enoch Bartlett, of Dorchester, near Boston (U.S.), introduced it into America, and now it is cultivated so widely that it is on sale for three or four months in the year, and exported also to England. Seckle, a good October pear, but small, we have from the States; the original tree is said to be near Philadelphia, about 100 years old. Clapp's Favourite (August) comes from Dorchester, Massachusetts; Dana's Hovey, "a veritable sweetmeat" (November and January), also comes from the same State. It is sometimes called Winter Seckle. Most of our good sorts are grown in the U.S., and Californian pears are now coming to us in great quantities. They are sent in wooden boxes, properly graded and packed. Every fruit is in paper, with the name of the grower on it, and the name of the variety on each box. The excellent quality and careful packing ensure a good demand at a high price. Good American sorts are Lawson or Comet, Block's Acme, Sugar Pear, Bloodgood, and others. Our growers may learn a useful lesson from Californian pears in the London market.

NOTES ON VARIETIES

Emile d'Heyst is said to be equal to Marie Louise in quality, to be hardier, and to be a better bearer. It is not a grand grower on the Quince, nor does the fruit keep long (October, November).

Althorp Crassanne is often a first-rate pear. Mr Knight (very eminent a century ago) called it the best of all. It lasts from October to December. The tree is hardy, and a good bearer, but the fruit is hardly large enough for exhibition.

Brockworth Park, almost identical with Bonne d'Ezée, was once a pear of great repute, being large and showy, but the flesh is coarse (November).

B. Bosc is largely grown in Kent as a market pear. It succeeds on a chalky, warm soil. It is sometimes "first rate," Barron (October, November).

Beurré Mortillet (new) (D. G., *i.e.* Double Grafted) is a large and handsome September pear; gather before it is ripe.

Conference (Rivers), comparatively new, is large, handsome, and a good bearer, but not first rate (November).

Fondante de Thirriott, or Thiriot (new), grows and bears freely, fruit large and good. "First quality," Barron (November and December).

Madame Treyve is a good September pear, red and yellow, in chalky soil. It bears freely, but is not first rate.

Bon Chrétien should be gathered gradually before it is ripe, and laid on the shelves. It is said that you must sit up all night to eat it just at the right time.

D. du Comice is regarded as the best all round Dessert pear grown. Marie Louise is tender and unreliable. Thompson's, some think, the best for flavour. It is smaller, and bears best on the Pear Stock.

Marie Benoist is recommended in many lists as a good late pear, but my experience has not been favourable. It is late and large when it bears.

General Todleben is large and handsome, but usually only fit for cooking (October and November).

Princess (new) is a late Louise Bonne, large and good; the tree bears well.

Beurré or Doyenné Sterkmans is a medium-sized, late pear (December, January, February), flushed bright red on one side; "second rate," Barron.

Beurré de Jonghe is a good Christmas Pear, but a slow grower, and needs a wall or orchard house.

Beurré Bachelier is large, handsome, a good bearer, but quite second rate (November).

Hacon's Incomparable is large and handsome, but second rate (November).

Swan's Egg was a popular pear fifty years ago for market, as the tree is hardy, bears well, and the fruit is good, but rather small (October).

Noveau Poiteau is a good exhibition pear, of vigorous growth, and bears well;

the fruit is excellent but does not keep well (November).

Pitmaston Duchess is an increasingly popular pear for market. It is very large, and on a cordon often handsome; in warm seasons of good quality, golden yellow when ripe. Bush trees on Quince bear well (October, November).

Duchesse d'Angoulême was a great favourite formerly, the tree growing and bearing well. Fruit often very large, but coarse and gritty. Crossed with Glou Morceau it has given us a child Pitmaston superior to the parents.

Josephine de Malines is pronounced by Mr Barron to be "always good." Hardy, and bears well on Quince (January-April).

Fouqueray is a large, good pear, an improved B. Bachelier (October).

B. Hardy is a great favourite with birds; they prefer and peck the best fruit.

B. Alexandre Lucas is large and handsome; pyriform, the tree is a good grower (October, November).

Triomphe de Vienne is a large and handsome September pear tree (D. G.), grows and bears well, comparatively new.

Marguerite Marillat, a very large, handsome September pear, bears well: comparatively new.

Michaelmas Nelis is a new variety, of which a specimen fruit has just been sent me by Messrs Bunyard. It is as delicious as the Winter Nelis pear (December and January).

RECEIPTS (FROM CASSELL'S "DICTIONARY OF COOKERY," SLIGHTLY ABRIDGED)

1. *To bake Pears.*—Rub half-a-dozen large hard pears with a soft cloth. Put them on a buttered baking tin into a slow oven, and let them bake gently for five or six hours. When tender, they are done enough, and are excellent if eaten with sugar. Probably cost 4d. Sufficient for three or four persons.

2. *Another way.*—Pare very smoothly a dozen large baking pears. Halve them, take out the cores, put them side by side into a well-brightened block-tin saucepan with a closely fitting cover. Pour over as much cold water as will cover them, add the thin rind of a small lemon, a tablespoonful of strained lemon juice, an inch of stick cinnamon, and fifteen grains of allspice. Put on cover, place the dish in a gentle oven, let it remain until the pears are tender, add a little white wine if liked. If such a saucepan is used, no cochineal will be needed. Time to bake six hours. Probable cost 1s. 8d. Sufficient for eight or ten persons.

To Preserve Pears.—Gather the pears before they are quite ripe, pare, halve, core and weigh them, put into a deep jar, allowing 3 lbs. of sugar to every 4 lbs. of pears, and just enough water to moisten the sugar, and to keep the fruit from burning. The strained juice and thinly-pared rind of a lemon and an inch of whole ginger may be put with every 2 lbs. of pears. Place the jar in a saucepan of boiling water, and let the fruit steam gently for six or seven hours. Turn it into jars, and at once fasten these down securely, and store in a dry, cool place. Two or three drops of cochineal added to the pears after they are cooked improve their appearance. Pears preserved thus will not probably keep good more than three or four months. Probable cost 8d. per lb.

Pears Preserved, Red.—If in preserving pears it is wished to give a deep pink tinge to the fruit and syrup, use a perfectly bright block-tin saucepan. If this is not convenient, add three or four drops of cochineal to the syrup or a small proportion of Red Currant or Red Gooseberry juice.

Pears Stewed.—Pare, core, and halve eight or ten good-sized pears, leaving on the stalks or not, according to taste; put them into a tinned saucepan, with 6 ozs. of loaf sugar, 6 cloves, 6 whole allspice, ¾ of a pint of water, and a glassful of port (?). Let them boil as gently as possible until quite soft but not broken. Lift them out, put them on a glass dish, and when the syrup is cold, strain it over them. Some cream or custard added is a great improvement. Time to stew the pears from two-and-a-half to three hours. Probable cost 1s. 4d. Sufficient for five or six persons.

For Compôte of Pears, Pears Frosted and Iced, Pears Pickled, and other such receipts, see same dictionary.

For another method of preserving, see plums.

To Preserve Pears (from an old author).—Pare them very thinly and simmer in a thin syrup; let them lie a day or two in the syrup. Make the syrup richer, and simmer again, and repeat this process till they are clear; then drain and dry them in the sun or a cool oven a very little time. They may be kept in syrup, which makes them more moist and rich, and dried as wanted. Jargonelles are said to be the best for this purpose.

To Bake Pears.—These need not be of a fine sort; but some taste better than others, and often those that are least fit to eat raw are best for baking. Wipe, but *do not* pare, and lay them on tin plates, and bake in a slow oven. When soft enough to bear pressure, flatten them with a silver spoon. When done thorough, put them on a dish. They should be baked three or four times, and very gently.

To Stew Pears.—Pare, halve or quarter large pears, according to their size; throw them into water, as the skin is taken off, before they are divided to prevent them turning black. Pack them round a block tin stewpan, and sprinkle as much sugar over as will make them pretty sweet; add lemon-peel, a clove or two, and some allspice cracked; just cover them with water, and add a little red wine. Cover them close and stew three or four hours; when tender, take them out, and strain the liquor over them.

PLUMS AND DAMSONS

WHAT IS THE FINEST FRUIT?

What is the finest fruit in the world? The secretary and the superintendent of the R.H.S. (in vol. xxvi., parts ii. and iii. of the *Journal of the R.H.S.*) agree in thinking that Goldoni, a yellow nectarine raised from a peach by the late Francis Rivers is, when properly ripened, without exception, the finest fruit in the world. It has not been my privilege to taste it, yet I venture to think that a thoroughly ripened plum of one of the best varieties must come near it. The incessant demand for greengages is a testimony to the popularity of the plum as a dessert fruit. Next to the apple, it is the most useful of our fruits.

ORIGIN OF THE PLUM

Eminent botanists are of opinion that our plums and damsons have had their origin in the *Prunus Communis* found in various parts of Europe and Asia, but others consider that the *Prunus Domestica* is the parent of the majority. Mr A. H. Pearson of Chilwell, Nott. (*v. Journal of the R.H.S.*, vol. xxi. part ii.), thinks that "the blood" of more than one species is found in the plums of the present day, as varieties closely resembling one another demand different stocks for their well-being when propagated by grafting. The cherry plum is *Prunus myrobalana*, and of this species there are several varieties, as St Etienne, Mirabelle Précoce, *i.e.* the Early Mirabelle, Mirabelle Petite, and others. Rivers' Early Prolific is said to be of the same race.

The Bullace is classed by some botanists under the *Prunus Instititia*, and they place the damson in the same species, but the latter is round, the former oval. The damson, a small plum, may be safely classed with the *Prunus Communis*. It derives its name from the city of Damascus. Damascena is the word used in Pliny for the district round Damascus, and damson originally meant the Damascus plum. The Chinese have for centuries cultivated plums, and in the United States plums from Japan are coming rapidly into use, and appear to be more successful there than in the British Isles. We find the word *prunum*, a plum, in Vergil, Ovid, Martial, and other Roman writers. *Prunus*, a plum tree, is derived directly from the Greek; *prunus silvestris*, in Columella and Pliny, is supposed to mean the black thorn or sloe tree. These illustrations prove that the plum has been known for ages, and that its value is recognised in every part of the world. Our word plum is plainly derived from the Latin (probably through the Anglo-Saxon), and the word prune is almost identical with *prunum*.

SOIL AND SITUATION

The plum is not so particular as the pear about soil, yet it has its preferences. It is not so deep-rooting as the apple and pear are; the character therefore of the lower soil is not so important. But stone-fruits require lime. In planting for profit, no site should be selected for a large plantation if the soil is deficient in lime. It is true that lime can be added, but this plan may suit a private garden, not a large plantation for profit. The plum being hardier than the pear will flourish in most soils, even in a heavy loam, but not in light sandy or gravelly soil. In the latter case, something may be done by heavy manuring and frequent removal. The trees in the R.H.S.'s garden at Chiswick are a triumph of skilful culture, as good crops are raised on many trees in a hot and gravelly soil. Some damsons, however, do not thrive there. But such culture is costly. In soils of an intermediate character, much may be done by adding other materials as suggested for pears. If there is any doubt about the amount of lime in the soil, an analysis should be obtained, and special notice taken of the trees in the neighbourhood. The plum (like the pear) will not thrive in a low, wet, undrained locality, nor in one that is very dry or exposed. Drainage is essential to success. If, in a rainy season, water in a clay soil is allowed to remain round the roots, canker or gumming is pretty sure to follow. Excessive moisture is as bad as extreme dryness. The slope of the ground, therefore, is a matter of importance. In Essex there is often land quite level with a heavy clay soil difficult to drain; such soil would not suit plums, though it might suit quinces. The aspect as well as the slope must also be considered. For the better class of plums, *i.e.* dessert varieties, where sweetness is expected, a position open to the southern sun is best, but they will also thrive if the aspect is S.-E. or even S.-W. Culinary and hardy varieties might be planted in the colder aspects to the N., N.-E. or N.-W. Proper shelter must by no means be forgotten. Bitter north winds may injure the bloom almost as much as frost or rain; strong winds from the E. or S.-W. may do great damage to heavy crops. Mr Lewis Castle in "Plums for Profit" (edited by myself, S.P.C.K.) suggests that "Canadian and Italian poplars make a good break if tall growers are required, but cherry plums, the myrobalan, will grow into a strong hedge in two or three years' time if the height be sufficient." Damson hedges serve a double purpose and afford good protection. He also suggests that some of the ornamental crabs are similarly useful for protection. Of these the Transcendant and Hyslop or Dartmouth produce good crops of lovely fruit which are excellent for cooking purposes and would probably sell well.

PROPAGATION AND STOCKS

The usual method of propagation is by budding and grafting. The stocks on which the different varieties are grafted are raised from stones. Mr Pearson states that six kinds of stocks are used in the best nurseries—*i.e.* the common plum, the Brussels, the Mussel, the Brompton, the Damas Noir or St Julien, and the Myrobalan. The secret of success is to work the stock with a variety which is of common parentage. Nearly all plums will grow upon the common plum stock, though some of them thrive much better upon other stocks. Prince Englebert and Diamond flourish upon Mussel, but not upon the Brompton. Belgian Purple will not grow upon either Brussels or common plum, but succeeds upon Damas Noir, Mussel, or Myrobalan. The accurate knowledge required points to the wisdom of purchasing trees only from nurserymen who make such trees a specialty.

The late Archdeacon Lea in his excellent book "Small Farms" dwells strongly on the folly of buying cheap stuff. Trees on unsuitable stocks or not true to name bring bitter disappointment after a few years. "Never purchase trees because they are cheap. Visit the nurseries, and pick out trees with clean healthy bark, even though they are smaller than others." If you cannot go or send a reliable man, write in good time and get an early choice. Select and accept only young trees not more than two or three years' old. Budded trees are better than those grafted, as a general rule, the union being better; indeed grafting is usually adopted because budding has failed. In trees that have been budded, there will probably be less gumming.

PLANTING

Planting is a matter of supreme importance, but the rules for pears and plums are very much the same. Especial care must be taken if the soil is heavy and loaded with moisture. Put the trees on arrival in a trench (see before), and wait until the ground is fit and the soil as fine as possible so that the roots may run freely through it. Get the stakes ready and place them in position before planting. Bind the tree, if tall, at once when planted to the stake by soft willow twigs or other means, taking every care that the bark is not rubbed by the stake. Old cloth or carpet may be used for this purpose, tarred twine or cord being passed round it. Dry stakes well tarred, often last as long as they are needed.

WHAT IS YOUR OBJECT?

What is your object? Before choosing varieties, or planting, it is advisable to ask yourself, what is my object? On the answer the form of plantation and the choice of trees must depend. If for a private house only, the answer is easy. Then comes the question, Is there a wall, and if so, what is the soil and the aspect? Is there an Orchard House? If for market, for what market are you preparing? In the Midlands, the Pershore (= Gisborne's) is a great favourite; in London, the Early Orleans and the Egg Plum; in the North, the Black Diamond, the Wydale and others. In planting damsons the same question should be put. The Midland people won't have the Farleigh Prolific so popular in Kent, and they are right; the Shropshire folks think their damson the best of all and many agree with them. Are you near a jam factory? What plums do they desire or require? Local circumstances and wants should have great weight. If you are near a wood and birds are numerous, you may be wise in not growing greengages, yet otherwise they may be the best sort for a large outlay as the demand for them is universal.

PLUMS FOR A PRIVATE GARDEN

Let us suppose that the soil is fairly good; the choice of trees is not difficult. We have a selection made in 1892 by a committee of the R.H.S., consisting of forty experts, and their choice has been confirmed in a remarkable degree by a report of the trial of plums at the Chiswick Garden of the R.H.S. in 1901. At this trial on a soil that in a good year is said to suit them generally, ninety-five varieties were tested, and a good account was given of the following ten:—

Plums for Eating.

Rivers' Early Transparent Gage.—"Green or greenish yellow, flushed with red, the finest early dessert plum, a good cropper, habit bushy, compact, vigorous."—R.[9] August 21.

Dennistoun's Superb Gage.—"Green and of greengage flavour, a first-rate dessert variety, of exquisite flavour, cropping well as a bush tree in the open air, habit erect, compact, vigorous."—R. August 26.

Jefferson's.—"Fruit larger than the two former, yellow, covered with small red dots, habit erect, compact, very vigorous, the best 'all round' table kind, succeeds in every form."—R. September 6.

In the list of 1892 *Belgian Purple* was named as a plum for eating, but it is only fit for the table in warm seasons, "dark purple, of medium size, bears well, habit erect, compact, vigorous."—R. August 19.

Plum—Rivers' Early Prolific

Plums for Cooking.

Rivers' Early Prolific, a dark purple fruit, rather small but one which comes in early so that it is often first in the market; thus the tree has time to rest and recover before winter. "A very early and valuable cooking plum; of fair quality for dessert, a great and constant bearer." The tree does not thrive everywhere, nor is it very vigorous.—R. July 23.

Rivers' Czar.—"Dark red or purple, of medium size, very good bearer, habit erect, compact, vigorous."—R. August 2.

Victoria, "Fruit large and bright red, very heavy cropper, the most popular plum and best for general purposes."—R. August 22.

Cox's Emperor, or Denbigh, or Denbigh Seedling, "a large dark red of the Orleans type, habit erect, compact, vigorous, a very good free bearing plum that al-

[9] The dates refer to the time when the fruits were "ready" (ripe, fit for gathering) at the Chiswick Garden of the R.H.S.

ways cooks well."—R. August 22.

Plum—Czar

Gisborne's (like Pershore), "Fruit deep yellow, rather large, habit erect, compact, vigorous, a great bearer."—Early September.

Rivers' Monarch, "Fruit large, bluish-purple, the best late C. (Cooking) plum, habit rather diffuse and vigorous."—R. September 13.

These ten varieties have borne the test of time, and won approval from the R.H.S. in 1892 and 1901. The descriptions are those of the R.H.S. Mr Lewis Castle omits Cox's Emperor and adds;—

Early Orleans, "medium size, reddish purple, good bearer and good C. variety, habit diffuse and moderately vigorous."—R. August 2.

Greengage.—If this is planted, choose July greengage, "rather large for a greengage, habit erect, compact, vigorous, better bearer than old G., fine-flavoured Dessert plum."—R. August 7.

Diamond, "very large, blue-black, very heavy cropper, habit bushy, compact, vigorous, good C. plum."—R. August 23.

Pond's Seedling, "Fruit very large, deep red, habit bushy, compact, vigorous, rather late, free bearing C. plum."—R. September 7.

Prince Englebert, "Rather large, dark purple, habit erect, compact, vigorous, reliable C. plum."—R. August 13.

Coe's Golden Drop, "Fruit large, pale yellow, habit bushy, compact, vigorous, a delicious late D. plum, an indifferent bearer on bush trees, most reliable on walls."—R. September 13.

These remarks are based on the Report of trials at Chiswick in 1901.

If one plum only is wanted, choose Victoria, if three, Early Prolific, Victoria and Monarch; to these Dennistoun's Superb and Jefferson might be added for dessert if five are desired.

All these varieties may be planted as Standards, but are better as half-standards or bushes. If as cordons they must be two feet or more apart, and lifted about once in three years. There is no dwarfing plum stock like the paradise for apples or the quince for pears.

If the dessert varieties are on walls, special attention must be given to pruning and to root-pruning. The growth of coarse thick stems and branches is often fatal to the prospect of good crops.

PRUNING AND TRAINING

Plums as a rule do not need nor will they bear as much close pruning as pears and apples. But they need special attention in early life.

STANDARDS.—These must be planted, when other crops are to be grown beneath. Quincunx fashion :·: is the best. The rows, as a rule, should be 24 feet apart, and the trees in each row about 20 feet. Plums do not shade as much as apples and pears, yet it is always wise to avoid overcrowding. Some sorts are not as spreading or as vigorous as others. Weak growers like the Early Prolific might be placed between Jefferson and Monarch. Good trees in six or seven years should bring good returns, but the intermediate space may meanwhile be utilised for strawberries, gooseberries, and so on. Standards should be 6 feet high or more. After planting, cut back the shoots to about one-third of their length, the weakest still more, to promote vigorous growth, and cut just above an outer eye. Keep the centre open. In later years stop gross or robber shoots in June, clipping some leaves of the latter, if necessary. Never allow boughs to cross, and keep all the tree fairly open. When the tree begins to bear, little pruning is necessary. But stopping luxuriant shoots about midsummer is good for the other branches, and for the production of fruit buds before winter. Complete pruning early in August. In winter cut out dead wood, and shorten boughs wherever fresh wood is required. The wood of the Victoria plum is very brittle, and requires special treatment. Shorten the strong growing luxuriant branches of this variety in July; otherwise later on they will break when loaded with fruit. Messrs Bunyard's choice of six for market standards is: Rivers' Early Prolific, Czar, Early Orleans, Victoria, Pond's Seedling, Rivers' Monarch.

HALF STANDARDS (3½ feet to 4 feet high) are better, and more manageable. Planted 12 feet apart, gooseberries, etc., may be placed around them; otherwise they may be nearer, even up to 6 or 8 feet. These should be pruned in August unless strong shoots require pinching back. Stop new side shoots at the sixth leaf to produce fruit-buds. Avoid excess. Wounds made in August have a better chance of healing while the sap is still active. Pyramids are not as useful as bushes; the former require a central stem and special training.

IN BUSHES, keep the tree open, stop strong shoots at midsummer, prune new (side) shoots back to six leaves about mid-August, and take out wood that is not wanted, admitting the sun and air. In winter cut back any boughs where fresh shoots are wanted to a wood bud at an outer eye.

TREES ON WALLS.—Plums are usually put on east walls, but the best repay a south as well as a west aspect. They require and repay care and skill in training. If the wall is low, the horizontal form is best. The branches should be taken several inches below the line along which they are to be trained, and not at right angles; the sap will flow better, and the tendency of branches to die off will be lessened. The first branch should be 1 foot from the ground, the rest 9 inches apart. Coarse stems and branches must be avoided by moderate root-pruning. The wood must be kept near the wall, that wood and fruit may be better ripened. The fan system is better for a high wall. Train shoots on the tree from the nursery in regular order at equal intervals, cutting back only to ripe wood. Pick off growths on the side next the wall, and others badly placed. Lay in new wood every year, and in August or Early September cut out unsightly branches or spurs if there is other wood to replace them. Prune upper part of tree first, and encourage foliage and fruit spurs over every part. Stop strong growing branches at midsummer, and pinch back side shoots to six leaves about mid-August. Fruit buds will follow. Wire on the wall should be 1½ inch out, with an interval of 1 foot between each wire.

MANURES

The absence of moisture and the consequent exhaustion have a serious effect on plums, and should never be allowed. Mulch newly-planted trees in light or poor soil; give liquid manure or irrigate in dry weather. Should the crops be heavy, and the soil at all deficient in lime, the deficiency should be made up by scattering lime some distance around the stems and working it gently in. "An annual dressing of decayed vegetable matter, old manure and lime-rubbish, laid about a yard round the stems, produces very satisfactory results" ("Plums for profit"). Supply this in November. The artificials recommended by R.H.S. for pears are also good for plums. Dr Griffiths recommends cow dung and a mixed manure, composed of 5 parts of kainit, 1 part of magnesium sulphate, 2 parts of superphosphate; 7 lbs. of the mixture to be applied to each tree in autumn, two more pounds in the spring.

Established trees in full health need no aid in an ordinary season if they carry no crop. Damsons should be fed as well as plums.

THINNING

must be carried out severely if the crop is thick. In a good plum season, only very fine and first-rate fruits fetch a good price, and these can only be obtained by thinning the fruit and feeding the trees. An annual crop (if frost does not interfere) may then be expected. Half the crop in some years should be taken off long before the fruit is ripe. The jam-makers utilise green and half-ripe plums.

GAGES

These are dessert plums, some of the highest excellence, but they usually require a wall or the best soil and situation. The demand for them is very great. Preserved with proper care, they last until plums come again, and often fetch a higher price than the red or dark plums.

The following are the five best:—

1. July Greengage (see before) early in August.

2. Dennistoun's Superb (see before) mid-August.

3. Early Transparent (see before) end of August.

4. Reine Claude Comte d'Althann, briefly Comte d'Althann, a comparatively new plum from Bohemia. First-class certificate R.H.S. "Medium to large, greyish green, deeply flushed and dotted with red, covered with a beautiful white bloom, very heavy crop, habit bushy, compact, vigorous, remarkably good dessert plum, succeeding equally well as a wall tree, bush, or standard, remarkably prolific." R.H.S., R. August 22.

5. Jefferson (see before).

6. If more are wanted M'Laughlin's Gage is "rather large, pale yellow, flushed with red, a good cropper, habit erect, compact, vigorous, one of the finest dessert plums." R. August 17.

7. Guthrie's Late Green, "very good cropper, habit bushy, compact, vigorous, a most delicious dessert plum." R. August 30.

Nos. 2, 5 and 6 are splendid specimens of American plums, No. 7 is Scotch, named after the late Mr Guthrie of Dundee.

The following are well known but not so good in some respects: Bryanston Gage, Oullin's Golden Gage, Golden Transparent. Coe's Golden Drop has been described. Angelina Burdett is sometimes classed with the gages. It is "dark red, shaded with purple, a good cropper, habit bushy and compact, a delicious dessert variety that hangs well in the trees for some time after it is ripe." R. August 22.

MARKET PLUMS

If a planter prefers to grow gages, he must protect his trees from bullfinches and other birds. The former often carry off the buds in winter, and ruin all hopes of a crop. Such a plantation near a wood would usually be a failure. If the trees are washed in early winter with No. II. mixture, the buds will have some protection. Lime should be thrown over the branches on a damp day. The gun in many cases must be at work from dawn to dusk. The gardener must learn to distinguish between friend and foe. Mr Lea's list in "Small Farms in the Midlands" is as follows: Early Prolific, Victoria, Black Diamond ("the wood is remarkably tough"), Pond's Seedling ("tolerably tough"), Pershore Egg Plum, *i.e.* Gisborne ("hardiest of all plums, surest cropper, comes early into bearing, the wood tough, and though the price is low, pays well"). He also mentions Prince Englebert and Jemmy Moore ("alias Cox's Emperor, alias Denbigh"), but wisely adds, these come in about the same time as Victoria, when there is a glut. Early or late varieties usually sell best. A new variety, Bittern, raised (as so many varieties have been) at Sawbridgeworth, by the late Francis Rivers, seems well worth trying: "Fruit rather large, deep purple, very heavy crop, habit bushy, compact, vigorous, excellent early free-bearing variety. R. August 2."

Curlew is another "early free-cropping purple plum, habit erect, vigorous," also raised by Francis Rivers. Monarch is a late, good, and very saleable plum. It is said that 75 per cent. or more of the plums planted in recent years have been Victorias. Planters should avoid the Victoria glut. Pond's Seedling, red and very large, coming in after Victoria, often sells well. Put up in a small basket it is inviting, and sells for an eating though properly a cooking plum. Early Prolific is also largely planted, but does not suit all soils. The white Magnum Bonum or egg plum (very large), good for cooking and preserving, sells well in London, but needs a strong soil. It is an early September fruit. Wyedale is popular in Yorkshire, valuable for its lateness, and for keeping sound on the trees when ripe; its habit is diffuse, size rather small, a good cropper. Mr Pearson names Goliath, a large reddish purple plum, "a good cropper, habit diffuse and vigorous, a very useful cooking plum. R. September 7."

At the Plum Congress held at Edinburgh in September 1889 an election of sorts was made with the following result:—

Dessert Plums

	Votes.
Jefferson	72
Coe's Golden Drop	69
Green Gage	57
Kirke's	57
Early Transparent	25
Oullin's Golden Gage	16
Reine Claude de Bavay	14

Culinary

	Votes.
Victoria	74
Pond's Seedling	54
White Magnum Bonum	52
Early Prolific	40
Goliath	31
Early Orleans	27

The friend who sent me this list remarks: "Newer plums supersede some of these. Czar does not crack with the wet as Orleans does. I prefer Diamond to Goliath." Kirke's is better on a wall than in the open. The same may be said of Oullin's; neither are grand croppers. Reine Claude de Bavay is a late gage, richly flavoured, but not a good cropper.

GATHERING, PACKING, MARKETING

A step ladder or a Heathman's combined ladder is best for gathering as plum wood is often brittle. Look over the trees several times and gather gradually. Fruit for home use should not be gathered until it is ripe, but for market it should not be quite ripe. Early morning when the fruit is cool is the best time. Dessert fruits generally should be handled as little as possible, otherwise the bloom on them and the appearance are spoilt. Plums are often sent away in round baskets, or oblong flat baskets. The former in the London markets are termed sieves or half-sieves. A sieve holds seven imperial gallons; the diameter is 15 inches, the depth 8 inches. Flat baskets with lids protect the fruit from injury. Stout and strong paper, above, below and around, assist in saving it. Oblong baskets with handles and without a lid are used in the Midlands and the N. They are called "pots," and local inquiry as to weight should be made. Strong brown paper is useful on the top. The cost per ream is from 10s. to 20s. But non-returnable boxes are better. The baskets are often missing or lost. The sizes of unplaned boxes with lids to be nailed on are usually as follows:—[10]

lbs.	Length.	Width.	Depth.	Gross.		Doz.	
				s.	d.	s.	d.
14	10¾ in.	10½ in.	6¼ in.	26	6	2	6
28	22 in.	10¾ in.	6¼ in.	46	0	4	2
42	27 in.	15 in.	8 in.	84	0	7	6

—GLOVER.

Choice dessert plums sent in light boxes (one or two layers only in a box) placed in wood-wool, and with tissue paper covering the inside of the box, and lying between each layer, often sell well. White paper-lace (such as is used for honey sections) sets off good fruit, and makes it more attractive. Pink tissue paper is often used for light fruits. The boxes should be uniform in size and quality of fruit as far as possible. Tissue paper can be bought at 2s. 6d. to 5s. or more a ream, and should not be grudged. The best wood-wool ranges from 18s. to 25s. per cwt. A few visits to Covent Garden, the best shops, and the Crystal Palace Fruit Shows, will not be time or labour lost.

Plums and damsons for market should be gathered and sent before they are quite ripe; if soft and pulpy on arrival, they are valueless. Sort in size and quality as even as possible: keep back all inferior stuff. Only good produce, well sorted and properly packed, placed on the market in good condition, is likely to sell well. Foreigners as well as neighbours compete for custom. In large establishments a packing room with every convenience close at hand is necessary.[11] The market-agent should daily advise what goods are needed.

STORING AND KEEPING

Plums, as a rule, do not remain good for any length of time after being gathered. They will however last a week or two if laid out in a cool, dark, well constructed place. Slate slabs assist to keep baskets and fruit cool. Some of the late dessert varieties gathered before they are quite ripe, wrapped in paper, will last in a dry place for a long period. Dr Hogg says that "Ickworth Impératrice," a large late dessert variety, if allowed to remain on the tree until it shrivels, then wrapt

[10] From Watson, vol. v. p. 369.

[11] For many useful details see Watson, vol. v.

in silk paper and placed in a dry shelf, will last for many weeks. It is a richly-flavoured plum. The remark is probably true of other late varieties; *e.g.*, the lovely Golden Transparent, "a delicious plum grown against a wall, but not a success in the open" (R. September 12), or Reine Claude de Bavay, which is late, but a poor bearer. The Ickworth Impératrice was not tested by the R.H.S., and is not now often grown. Guthrie's late Green, "a most delicious dessert plum and the heaviest cropper here (*i.e.* Chiswick R.H.S.) of all the gages," is probably one of the best sorts for keeping as described above. Angelina Burdett (see gages) "if allowed to hang till it shrivels becomes a perfect sweetmeat" (Hogg).

INSECT ENEMIES

1. Aphides are often a great trouble. There are three sorts or more, one called the plum aphis. They attack in spring and cause the leaves to curl up, and so check growth. Steep 4 ozs. of quassia chips in a gallon of soft water for twenty-four hours. Dissolve 2 ozs. of soft soap in this mixture, and add to the infusion. Apply by a painter's brush, and carefully wash the under side of the leaves (Rivers). On a larger scale: "Boil 1 lb. of chips in a gallon of water for twenty minutes, strain off the chips and add 38 gallons of water. Put 1 lb. of soft soap in a gallon of water until dissolved, then add to the rest. Apply by a syringe or sprayer. Where the foliage is young follow the spraying by syringing half-an-hour afterwards" (Castle). Tobacco water made as follows is also a good remedy: "Pour soft boiling water at the rate 1 gallon to 2 ozs. of the strongest shag tobacco, allow it to stand till cool. Its efficacy is increased by dissolving 2 ozs. of soft soap in each gallon at the time it is poured on the tobacco" (Wright). This mixture may be applied with some force by the garden engine. The great point is to syringe or paint with one of these remedies as soon as the evil is perceived.

2. Red Grub is often very injurious. The moth measures about ½ inch across, the caterpillars are pale red, with brown neck and black head. They pierce and drop with the fruit, seek shelter in the bark, where they spin a cocoon and pass the winter. If the trees have been scraped, then washed with a mixture of lime and soot, paraffin and grease (see No. III. pears), or sprayed before the buds open with Bordeaux mixture (see No. II.), and also afterwards, they probably will not suffer. Lime and soot scattered over the ground under the trees will also be useful. If the plums are attacked, collect all fallen fruits and shake the trees every morning, burn the fruits affected or give them to the pigs.

3. The Plum Sawfly also attacks the fruit, laying an egg in the calyx of each flower. The grub is whitish, with brown head. It enters the fruits, feeds on the stone, and causes them to drop. A spraying of the modified mixture No. II. after the fruit has set would be useful, but as the grub pupates in the soil, lime and soot will again be serviceable. Collect and destroy fallen fruit daily.

4. Red Spider, a spinning mite, is a great pest in dry summers. It must be checked by the free use of the syringe or water engine as soon as seen. Yellow spots on the leaves are a proof of its presence. Mix 4 gallons of soft soap solution with ½ lb. of flowers of sulphur; apply with syringe. Strong soap-suds, or even clear water forcibly given are better than nothing.

Fungoid attacks injure the trees. The Bordeaux mixture (No. II.) is the best preventive and remedy if there are any signs of fungus. Cut away all diseased twigs, boughs and branches, and burn them. Fungus spores are scattered by the wind and spread the disease. Drench the trunk and bark in winter with this mixture before the buds swell. Care must be taken not to apply the mixture in full strength to tender leaves and buds.

For the fungus mildew, half an ounce of sulphide of potassium mixed in a gallon of water and applied by a syringe is recommended (Wright). Finely-powdered quick-lime mixed with sulphur (double the quantity of the former), and distributed by a special bellows (see before, page 23), is also said to be a good remedy.

ORCHARD HOUSE

For dimensions see under pears. Plums are best in pots or tubs, as they can be taken out when at rest. They are very liable to attacks from aphides, but the insecticide for pears in pots is good also for plums. The house must be fumigated, and the trees syringed on the least appearance of aphis. Place the pots on bricks (*v.* pears). When growth is being started the temperature should be from 45° at night to 50° by day. Soft or tepid water should be given freely. Fumigate again just before the flowers come out. As the buds increase, raise the temperature 5° to 10° and syringe once or twice a day with tepid water. But a dry atmosphere is important while the trees are in flower. Admit air as well as bees in the forenoon, and pass a camel-hair or light brush over the flowers about the middle of the day. When the fruit is set, syringe at least once a day; if the weather is hot, twice or even three times a day, and give all the air possible. Thin the fruits (if the crop is large) with scissors; mulch and feed with weak liquid manure (see pears). The shoots must be pinched if the trees are of any age, at the fifth or sixth leaf. Not much heat is needed generally, but when the stoning period is passed, the ripening process may be hastened by a higher temperature. The house may be closed at an earlier hour if necessary. Avoid extremes. As the fruits ripen, cease gradually to syringe, but keep the house moist by sprinkling water over the paths, etc. Choose the choicest dessert sorts: Early Transparent, Dennistoun's Superb, July Greengage, Jefferson, Count Althann, Coe's Golden Drop, Guthrie's Late Green, Angelina Burdett, Bryanstone Gage, and Golden Transparent; and if darker colours are desired: Early Prolific, Belgian Purple and Czar. Bryanston Gage was recommended by the R.H.S. in 1892, and is a very richly-flavoured dessert variety, but is not a good cropper in the open, and needs a wall or house.

DAMSONS

are often very valuable, and also make good outside hedges. Bradley's "King of the Damsons" is the best. The fruit is large, the tree "free-cropping, bushy, vigorous, erect." R. September 20. Frogmore Prolific (earlier) is also large and free-bearing. R. September 9. Both these are late. Mirabelle and Rivers' Early Damson are August damsons, small, the former vigorous.

BULLACES

Shepherd's is the best, and hangs late on the tree. A few trees in a large garden are useful. R. September 20.

IMPORTANT POINTS

Good sorts on suitable stocks in good soil and proper aspect; lime in the soil, added or otherwise; winter washing or spraying; thinning fruit; early training; moderate pruning; root-pruning in very strong soils; lifting in shallow soils; liquid and other manures; immediate action if aphides or red spider appear.

DRYING BY EVAPORATION

This important subject cannot be treated here at length. In a hot season with abundant crops, good results may be obtained with some prospect of profit. But the apparatus has been expensive. Mr Udale's Report to the Worcestershire *C.C.* on dried fruits, vegetables and herbs, with the article in *Journal of R.H.S.*, vol. xxvi., part ii., should be consulted, and "Fruit Preserving," by R.L.C., in Watson, vol. v. Thick-skinned plums, *e.g.*, Czar, Prince Englebert, Diamond and Monarch are best for the purpose. Plums placed on trays, dried in a very slow oven, and allowed to cool several times, are often equal to French prunes.

BOTTLING

This is a simple and most useful process. Plums well bottled will last for years. Gather clean and dry fruit before it is quite ripe, that the heat may not crack the plums. Remove the stalks and pack closely in bottles not over 11 inches high, without bruising, up to shoulder of bottle. Provide a boiler a foot deep; place hay or canvas at the bottom, then put the bottles in the boiler with hay or canvas around them to prevent fracture. Now fill the boiler up to the necks of the bottles, and place it on a slow fire. Heat very gradually until the water is at boiling point. Then take each bottle out with a cloth, fill with boiling water kept close at hand, and cover *while boiling* with air-tight stoppers. Another method is to fill the bottles nearly full with cold water or thin syrup, and boil for fifteen minutes. Messrs De Luca have received silver and bronze medals from the R.H.S. for self-closing bottles now sold by Messrs Abbott of Southall, near London. Their method is as follows: "Pour in water or cold thin syrup (one tablespoonful of crystalised cane-sugar to the pint) sufficient to cover the fruit. Adjust the indiarubber in the groove made for it on neck of the bottle, place the disc on it, and *lightly* screw down the outer ring. (Steam must be allowed to escape.) Boil as before for twenty minutes; take out each bottle, and at once screw the outer ring as tightly as possible. Leave bottles until cold. Next day examine by unscrewing the outer ring, and try whether the disc is firmly fastened down. If so, replace the ring, screw down tightly and store away in a cool place, standing them upright. The bottles by having new discs and indiarubber rings may be frequently used."

The Rev. W. Wilks, secretary of the R.H.S., recommends pears, especially Pitmaston D., as suitable for bottling. "Bottled it is delicious." He thinks fifteen minutes from the time the water boils sufficient for plums. Messrs De Luca mention an hour as the time for pears.

Messrs Lee & Co. of 19 Knightrider St., Maidstone, have received medals from the R.H.S. and others for their fruit bottling apparatus and bottled fruit.

They supply a patent economic fruit bottling apparatus at 21s. A thermometer at the side records the temperature of the bottles and of their contents. The following is the method given in the *Journal of the R.H.S.*

"The fruit must, of course, be fresh and good and the bottles clean. The fruit is first packed into the bottles, which are then filled up to the neck with cold water, or if preferred, with thin syrup made by dissolving ½ lb. of loaf-sugar in hot water and leaving it to cool. The bottles are then put into the cooking pot where they must remain for certainly not less than two hours at a temperature of between a minimum of 145 degrees and a maximum of 160 degrees. This low sterilising temperature being maintained for two, three, or four hours will destroy all germs without cooking the fruit, and is the great secret of successful bottling. No actual harm is done by the heat rising above 160 degrees, but if it does the fruit will probably burst, lose its shape, and not look so nice. Vegetables may be preserved in exactly the same way, but they must be done twice over with an interval of twenty-four hours to allow of their becoming quite cold. Jams and fruit jellies can be preserved fresh and good for years in the same way."

Success in bottling and preserving fruit depends chiefly on two points: (1) The destruction of every germ of mildew, etc., by keeping the bottles at a certain temperature for a certain time; (2) the absolute prevention of any possible re-entry of air into the bottles afterwards. The bottles must be hermetically sealed while in the steam or standing in almost boiling water (see *Journal R.H.S.*, vol. xxvi. part iii. p. 365).

Bottling or Canning in Syrup.—This is done by boiling together at the rate of 3 lbs. of cane-sugar to 1 quart of water and the white of 1 egg; pour the fruit whole into the syrup while boiling, and continue to boil together for only a few minutes, then pour into bottles or cans, and stopper or seal air-tight immediately whilst boiling. Pears may be preserved in the same way. Cheal, *Journal of R.H.S.*, vol. xxi. 1.

Plum Jam makes a rich preserve. Take equal quantities of fruit and cane-sugar; boil quickly half to three-quarters of an hour, then put into hot jars and cover well at once. Exclude the air as much as possible. The colour of the flesh is said to make a difference in the sale. Red jam is usually preferred, but greengage is also popular. Coe's Golden Drop or Autumn Compôte (September, hardy, fairly productive, but the fruit often splits) are good for yellow ("Amber"—Dr Hogg) jam; Belle de Septembre (September, "a good late cooking variety," good bearer, but fruit prone to split) for crimson colour. Free-stones are better than cling-stones. The following are free-stones: Bittern, Coe's Late Red ("fruit small, good bearer, a very useful late cooking plum"), Early Orleans, Early Transparent Gage, Old Greengage, Orleans, Oullin's Golden Gage, Red Magnum Bonum, Comte d'Althann, Victoria. The following are nearly so: Early Prolific, Czar, Cox's Emperor, Jefferson.

Belle de Septembre is a cling-stone. Damsons make good jam, the colour being a dark rich red.

PLUM JELLY

Plums are rich in "vegetable jelly." 1. Boil alone for half-an-hour, then strain the juice through a fine sieve or cloth; add 2 lbs. of cane-sugar to each quart of juice, boil again for twenty minutes, pour into jars and glasses, cover at once. A firm, clear and bright jelly should result (Watson).

2. "During the preserving season I generally have a few pots of jelly made from each pan of preserves without spoiling my jams. I make currant, gooseberry, and plum jelly this way.

"For all common preserves I allow ¾ of a pound of loaf-sugar to each pound of fruit. The sugar must be broken small. Put the fruit and sugar into your pan, let the sugar melt, then boil quickly for twenty to thirty minutes. Skim carefully, take the pan off the fire, take from it three or four cupfuls of juice, or as much as you think can be spared without making your jam dry. Strain the juice through a small gravy sieve into small jars. This will be found to jelly well. In this way a nice stock of jelly can be procured, and no fruit is wasted." (From Weldon's "Menu Cookery Book," 1s., published by Weldon, 31 Southampton Street, Strand.)

CHERRIES

"Keeping" Fruit, Suitable Soil, Aspect, Sorts Recommended by R.H.S. and Mr Bunyard for Eating, Cooking and Market, Protection, Pruning, Training, Cherries on Walls, Insect Attacks, Distances, Manures, Marketing, Derivation of the Word

It is useless to plant cherries unless the fruit can be protected from the birds. The cost of "keeping" a few trees would absorb all profit. In planting for sale, select two or three varieties only; and these should come in together, if possible, to lessen the cost of "keep." They should be intermingled, for reasons already mentioned (see pears, p. 7).

Cherries like a deep, mellow, and rather sandy soil, but they also thrive on a good loam lying on chalk. Stiff moist soils and dry gravelly soils are not suitable. The trees require much moisture, especially sorts with large leaves, such as the Bigarreau and Heart Cherries. Plant varieties to suit the soil. Inquire carefully what sorts do best in your neighbourhood. Cherries do well in open ground, not shaded nor in a valley. They prefer a south aspect, but Morellos thrive on a north wall. Kentish and Late Duke might also be tried there. Plant as you would pears or plums. Protect your trees from rabbits by wire, and from cattle by "cradles," 6 feet high at least,[12] or iron guards. Cattle fed on cake are useful in cherry orchards, and improve the produce.

Cherries for Eating, recommended by R.H.S. in 1892, are:—

a. Dukes.—(1) *May Duke,* dark red and early; (2) *Archduke* (large blackish red), mid-season, both tender-skinned, and so beloved by birds. Both are close growers.

b. Yellow-fruited (*i.e.* White Heart or Bigarreau).—(1) *Elton,* very early, good on heavy soil, tree spreading; (2) *Kent Bigarreau,* early, large, hardy, makes a large tree, best for general crop.

c. Black-fruited.—(1) *Black Eagle,* very large, travels well, free grower, mid-season; (2) *Early Rivers,* first-rate, earliest black, a good cropper and grower.

Cooking Cherries.—(1) *Kentish,* bright red, mid-season, a compact grower, best for jam; (2) *Morello,* very late, only fit for wall, fence, or bush.

Other good sorts are *Governor Wood* (mottled red, early) and *Black Tartarian,* a strong grower, fruit large, late. *Florence* is very large and late, succeeding the Bigarreau, but requires a warm soil and aspect.

---

[12] Cradles in Kent are often made of chestnut wood split, and last as long as they are needed. For form see pears, p. 4.

A Cherry Orchard

Where there is no wall, Morellos and other sorts should be planted as bushes in a garden 5 or 6 feet apart. They should be covered with nets when the fruit is colouring. Morellos last a long while on a north wall, protected by a net. These are often in great demand, and in a plantation succeed as standards. But the cost of "keeping" for a long time would be excessive.

PRUNING AND TRAINING

Allow the trees to grow a year before pruning them. Then cut back the branches to about 1 foot in length to an outer eye; the Bigarreau to 15 inches. Encourage two shoots from each branch, one at the end, the other 3 inches from the stem, and on the outside. Thus the branches will not grow into each other. Maintain the chief branches at nearly equal length for some years. Standards do not require much pruning. Cut back in summer (July) all shoots crowding the tree or threatening to cross others. In winter look over the trees again. Keep the trees open to sun and air. Cherries on wall should be on the fan system, and pinched back in July. The branches should be 9 inches apart for Duke Cherries, 12 for Bigarreau. If trained horizontally or as espaliers, shorten the shoots on the upper branches a week or two before those on the lower. Keep the shoots near the wall. When sufficient main branches are formed to cover the wall, do not cut back the leaders again. Be sparing with the knife. After two or three years fruit spurs will form, but not much surplus wood. Morellos require special treatment. The fruit is usually found on the wood of the previous year. Train young shoots in to replace the old, and cut out, when the fruit is off, all the old that can be spared. In standards, pyramids and bushes, thin out the branches.

Aphides are the chief enemy, especially on walls. Syringe with tobacco water made by pouring 1 gallon of soft boiling water on 2 ozs. of strong shag tobacco; add 2 ozs. of soft soap to the water when poured on the tobacco. Strain off and use cold. This solution is also good against the pear slug-worm, which attacks cherries as well as pears. Follow this prescription by a good syringing of cold water the following morning. The roots of cherries are near the surface so that the ground above them must not be dug.

STANDARDS

in strong soil should be 30 feet apart, 24 feet in lighter ground. They are best on grass which is used for pasture. The trees then should be 30 feet from each other. The Kentish Red are sometimes 15 or 18 feet only. Between standards at 24 feet apart bush trees of various kinds may be planted (apples, pears, plums), the two former on dwarfing stocks; there should be two between each standard 8 feet apart.

Ordinary manure is not often given. It may cause rank growth. Dr Griffiths recommends the following artificials: 3 parts weight of kainit, 2 parts of superphosphate, 1 part of nitrate of soda. Three lbs. of this mixture should be applied to each tree shortly before active growth begins. If the land is deficient in lime or chalk some should be given to each tree.

W. C. (in Watson): "Superphosphate of lime, 5 lbs., sulphate of potash, 2 lbs., sulphate of magnesia, ½ lb., chloride of soda, ½ lb. Apply during mild weather in February at the rate of 4 ozs. to the square yard of border, or the full quantity 8 lbs. to each rod of orchard ground."

Gather fruit dry before it is quite ripe. Cherries are usually sent to market in baskets which contain 24 lbs. nett; very choice fruit in 12 lb. baskets.

The word cherry comes from the old English cheri, chiri, and that probably from the French cerise, that from the Latin cerasus, and that from the Greek (κεϱασός) kerasos. «Cheri or chiri was a corruption of cheris or chiris, the final *s* being mistaken for the plural inflection; the same mistake occurs in several other words, notably in pea as shortened from pease, Latin pisum" (Skeat).

THE MULBERRY

ORIGIN OF THE WORD, SOIL, SITUATION, ASPECT, SHAPE, CULTURE, PRUNING, TREES IN POTS, USE OF FRUIT

The mulberry is a very handsome tree well worthy of cultivation in a large garden, if only it receives the care and culture which it deserves. Its proper name derived from the Latin through the Anglo-Saxon is Murberry. Mulberry is certainly more euphonious. It is said to be a native of Persia, but it has been known in this country for three centuries and a half at least. It is stated that there are trees still living among us several centuries old. The black mulberry is the one commonly grown in England; the white does better in a warmer clime, and has been largely planted in France and other lands, where the leaves are required for silkworms. The white, however, can be grown in this country. The mulberry will thrive in any good soil, but the ground should not be very wet. It should be placed in a sheltered southern aspect, and is likely to do well in the south rather than the north of England. It is propagated by layers, cuttings and seeds. Trees with good straight stems should be had from the nurserymen, and a few shillings spent in buying a straight, strong tree may save years. Plant in late autumn without manure with the usual care (see planting pears); the roots should not be exposed to frost or cold winds. Brick and lime-rubbish below in a heavy wet soil should be given. In a lighter soil put decayed manure round the tree after planting. In spring fork the soil up lightly, as trees thrive better when the soil is stirred. If you plant early in November under favourable circumstances leave four branches only, and cut these hard back to an outer eye. If you plant late, don't prune until the following winter. The first branches if cut back to three or four inches will probably throw out strong shoots. In August choose four of the best placed and strongest of these, cutting out the rest that those left may grow and the wood be ripened. The tree bears on short-jointed young wood, and on spurs, not on gross shoots. Keep the tree open, especially the middle, removing weak, watery spray, and train the shoots upwards. It will be a pleasure to go under the tree in time and enjoy its shade. The sun and air must have free access if the fruit is to ripen. Sometimes the berries fall prematurely. Drought or want of food may be the cause. Liquid manure in summer as the fruit comes on, and the artificial manures recommended for all fruit trees (see pears) will be helpful. In the winter-pruning, cut out all cross boughs, beginning from below, so as not to tear the bark. Check only branches that are rampant and running away from others. Keep the tree as uniformly round as you can. When the tree has (in four or five years) gained size and fruits well, grass may be grown around the tree, but it should be kept closely mown, especially when the fruit is ripening. All coarse grasses should be spudded out. If the fruit were thinned and the tree well fed, no doubt the berries would be finer. In any case feed well in July. The fruit is not very marketable, as it does not travel well, nor last long. But in cider counties it is sometimes mixed with apples, to make mulberry cider. The trees bear forcing in pots, and give good fruit in July. They will bear a high temperature. The fruit mixed with apples in a tart or pudding is described as "delicious." If it is gathered perfectly dry, it can be used to make a jelly in a similar manner to red currant jelly, and used for light puddings, etc. Mulberry syrup is said to be good for sore throat; mulberry water to be refreshing as a drink in cases of fever, mulberry vinegar to be efficacious for medicinal purposes just as raspberry vinegar, which it somewhat resembles. "Cassell's Dictionary of Cookery" explains these details, and also how to make mulberry preserve.

APPENDIX

THE PROPAGATION OF THE PEAR

It is not proposed to enter into elaborate details with reference to the propagation of the pear, for the general cultivator is far more concerned with and interested in its successful treatment when established than before, for he can now obtain prepared trees so readily and cheaply from nurserymen, who propagate them by the thousand, that it is not at all worth his while to attempt their propagation himself.

Pear trees are propagated either by budding or grafting. The first mentioned process is performed during July and August, and grafting is done in the month of March.

Budding consists in removing a bud from one tree and inserting it under the bark of the stock or branch of another tree. This work is done in the months of July and August, because the bark is during that time more easily raised, and a union more easily effected between the cambium (an inner growing layer) of both bud and stock. The buds inserted are taken from the current year's shoots, choosing shoots that are firm and short-jointed. After having removed a shoot, say nine or ten inches long, and cut the leaves to half their lengths, next proceed to cut out a bud. This is done by inserting a knife below the bud at a distance of about half an inch, and then drawing it upwards behind the bud, emerging again about half an inch above the bud. The cut must, of course, turn inwards—that is to say, towards the centre of the shoot. When removed, the bud therefore will be about an inch long, with the "eye" in the centre, and with a certain amount of wood attached to it behind. This should be removed, and the best way to do it is to insert the point of one's knife just underneath, so as slightly to raise the wood. Then, with the blade of the knife and one's thumb above, it can easily be removed with a slight jerk. Take great care that the root of the bud is not removed also. The stock to be budded should have a T-shaped incision made in the bark. With the ivory handle, which a proper budding knife will have, raise the bark on either side of the longitudinal slit, commencing at the corners just below where it joins the transverse incision. Take great care that the knife handle does not penetrate beneath the inner bark, but press it against the latter, slipping it along. When the bark is sufficiently raised, carefully insert the bud beneath, taking hold of it by the remaining portion of the leaf stalk. It must not be forced down, but introduced as gently as can be, otherwise there will be danger of injuring the vital cambium layer, where the union is effected. Afterwards tie the bud around with matting, to keep it in position and to prevent the entrance of air. Tie both above and below the "eye," leaving this of course free. An excellent indication as to whether or not the bud has taken is afforded by the petiole (leaf stalk). If this, a few weeks afterwards, falls completely away, one may be fairly certain that the bud has taken; if, on the other hand, it withers away, one may be almost equally sure that the operation has not been successful. The buds that have taken will commence to grow in the following spring, and then the stocks must be cut back to within a few inches of where the buds are inserted. Many nurserymen, if they find that the bud has not taken, graft the same stock in the following spring, instead of waiting to bud again in summer.

Grafting consists in so attaching one shoot to another that they unite and grow together. There are many different methods of grafting, but that most usually employed in the grafting of pear trees is tongue or splice grafting. This is done in the month of March, with firm growth of the preceding year. First cut the stock in a sloping direction, and so that the cut may terminate just above a bud if possible. "Great care must be taken that the scions fit the stocks," is the recommendation of one of our large nursery growers of pear trees, and one that should be closely followed. Therefore choose a graft as nearly as possible of the same size as the stock. Having cut the graft to a suitable length, say nine or ten inches, cut the lower end (that is, the thicker one) exactly as the top of the stock was cut, so that when placed together they will properly fit upon each other. The essential point is, of course, to make sure that the inner bark of the one coincides with the inner bark of the other, as then the cambium layers will also coincide. The graft should be carefully and securely tied on to the stock with raffia, and then covered over with either clay or grafting wax, so as effectually to prevent the entrance of air. It is advisable to place moss over the clay, so that in dry weather this can be watered to prevent the clay from becoming dry. Unless there were a covering of some material capable of holding moisture, it would be impossible to keep the clay in anything like a moist condition. When it is seen that the graft commences to grow freely, the clay may be partly removed so as to allow of the ligatures being loosened. It is wise to tie the graft to a stake attached to the stock when the clay is finally removed, as in rough weather it might possibly be broken off.

Lector House believes that a society develops through a two-fold approach of continuous learning and adaptation, which is derived from the study of classic literary works spread across the historic timeline of literature records. Therefore, we aim at reviving, repairing and redeveloping all those inaccessible or damaged but historically as well as culturally important literature across subjects so that the future generations may have an opportunity to study and learn from past works to embark upon a journey of creating a better future.

This book is a result of an effort made by Lector House towards making a contribution to the preservation and repair of original ancient works which might hold historical significance to the approach of continuous learning across subjects.

HAPPY READING & LEARNING!

LECTOR HOUSE LLP
E-MAIL: lectorpublishing@gmail.com

9 789393 693792